ENERGY SECTOR STANDARD OF THE PEOPLE' S REPUBLIC OF CHINA

中华人民共和国能源行业标准

Code for Design of Hydraulic Tunnel

水工隧洞设计规范

NB/T 10391-2020

Replace DL/T 5195-2004

Chief Development Department: China Renewable Energy Engineering Institute

Approval Department: National Energy Administration of the People's Republic of China

Implementation Date: February 1, 2021

China Water & Power Press

中国水利水电出版社

Beijing 2024

图书在版编目（CIP）数据

水工隧洞设计规范 : NB/T 10391-2020 = Code for Design of Hydraulic Tunnel（NB/T 10391-2020） : 英文 / 国家能源局发布. -- 北京 : 中国水利水电出版社, 2024. 8. -- ISBN 978-7-5226-2723-6

Ⅰ. TV672-65

中国国家版本馆CIP数据核字第202457RS94号

ENERGY SECTOR STANDARD
OF THE PEOPLE'S REPUBLIC OF CHINA
中华人民共和国能源行业标准

Code for Design of Hydraulic Tunnel

水工隧洞设计规范

NB/T 10391-2020

Replace DL/T 5195-2004

（英文版）

Issued by National Energy Administration of the People's Republic of China
国家能源局　发布
Translation organized by China Renewable Energy Engineering Institute
水电水利规划设计总院　组织翻译
Published by China Water & Power Press
中国水利水电出版社　出版发行
Tel: (+ 86 10) 68545888　68545874
sales@mwr.gov.cn
Account name: China Water & Power Press
Address: No.1, Yuyuantan Nanlu, Haidian District, Beijing 100038, China
http: //www.waterpub.com.cn
中国水利水电出版社微机排版中心　排版
北京中献拓方科技发展有限公司　印刷
184mm×260mm　16 开本　6 印张　190 千字
2024 年 8 月第 1 版　2024 年 8 月第 1 次印刷
Price（定价）: ¥ 995.00

Introduction

This English version is one of China's energy sector standard series in English. Its translation was organized by China Renewable Energy Engineering Institute authorized by National Energy Administration of the People's Republic of China in compliance with relevant procedures and stipulations. This English version was issued by National Energy Administration of the People's Republic of China in Announcement [2023] No. 5, dated October 11, 2023.

This version was translated from the Chinese Standard NB/T 10391-2020, *Code for Design of Hydraulic Tunnel*, published by China Water & Power Press. In the event of any discrepancy in the implementation, the Chinese version shall prevail.

Many thanks go to the staff from relevant standard development organizations and those who have provided generous assistance in the translation and review process.

For further improvement of the English version, any comments and suggestions are welcome and should be addressed to:

China Renewable Energy Engineering Institute
No. 2 Beixiaojie, Liupukang, Xicheng District, Beijing 100120, China
Website: www.creei.cn

Translating organization:

POWERCHINA Chengdu Engineering Corporation Limited

Translating staff:

YANG Huaide	ZU Wei	LIU Bin	DONG Ao
WANG Yongsheng	JIA Pan	ZHU Wenwei	ZHENG Yongjin
YANG Xingyi	LI Wanjun	YANG Minggang	DU Zhenyu
XIE Jinyuan	YIN Chonglin	PAN Liru	

Review panel members:

GUO Jie	POWERCHINA Beijing Engineering Corporation Limited
YAN Wenjun	Army Academy of Armored Forces, PLA
QI Wen	POWERCHINA Beijing Engineering Corporation Limited
YE Bin	POWERCHINA Huadong Engineering Corporation Limited

HAO Peng	POWERCHINA Guiyang Engineering Corporation Limited
PENG Fuping	POWERCHINA Kunming Engineering Corporation Limited
YANG Qing	Sichuan University
LI Zichang	Sichuan Energy Internet Research Institute, Tsinghua University

National Energy Administration of the People's Republic of China

翻译出版说明

本译本为国家能源局委托水电水利规划设计总院按照有关程序和规定，统一组织翻译的能源行业标准英文版系列译本之一。2023年10月11日，国家能源局以2023年第5号公告予以公布。

本译本是根据中国水利水电出版社出版的《水工隧洞设计规范》NB/T 10391—2020翻译的，著作权归国家能源局所有。在使用过程中，如出现异议，以中文版为准。

本译本在翻译和审核过程中，本标准编制单位及编制组有关成员给予了积极协助。

为不断提高本译本的质量，欢迎使用者提出意见和建议，并反馈给水电水利规划设计总院。

地址：北京市西城区六铺炕北小街2号
邮编：100120
网址：www.creei.cn

本译本翻译单位：中国电建集团成都勘测设计研究院有限公司

本译本翻译人员：杨怀德 祖威 刘斌 董傲 王勇胜 贾攀 朱文炜 郑咏珽 杨兴义 李万军 杨明刚 杜震宇 谢金元 尹崇林 潘礼儒

本译本审核人员：

郭洁 中国电建集团北京勘测设计研究院有限公司

闫文军 中国人民解放军陆军装甲兵学院

齐文 中国电建集团北京勘测设计研究院有限公司

叶彬 中国电建集团华东勘测设计研究院有限公司

郝鹏 中国电建集团贵阳勘测设计研究院有限公司

彭富平 中国电建集团昆明勘测设计研究院有限公司

杨庆 四川大学

李子昌 清华四川能源互联网研究院

国家能源局

Announcement of National Energy Administration of the People's Republic of China [2020] No. 5

National Energy Administration of the People's Republic of China has approved and issued 502 energy sector standards including *Technical Code for Real-Time Ecological Flow Monitoring Systems of Hydropower Projects* (Attachment 1), and the English version of 35 energy sector standards including *Series Parameters for Horizontal Hydraulic Hoist (Cylinder)* (Attachment 2).

Attachments: 1. Directory of Sector Standards
2. Directory of English Version of Sector Standards

National Energy Administration of the People's Republic of China
October 23, 2020

Attachment 1:

Directory of Sector Standards

Serial number	Standard No.	Title	Replaced standard No.	Adopted international standard No.	Approval date	Implementation date
…						
7	NB/T 10391-2020	Code for Design of Hydraulic Tunnel	DL/T 5195-2004		2020-10-23	2021-02-01
…						

Foreword

According to the requirements of Document GNKJ [2015] No. 12 issued by National Energy Administration of the People's Republic of China, "Notice on Releasing the Development and Revision Plan of the Second Batch of Energy Sector Standards in 2014", and after extensive investigation and research, summarization of practical experience, and wide solicitation of opinions, the drafting group has prepared this code.

The main technical contents of this code include: general provisions, terms, basic requirements, tunnel layout, shape and size of cross section, hydraulic design, unlined tunnel and anchor-shotcrete tunnel, basic principles for structural design, concrete and reinforced concrete lining, prestressed concrete lining, high-pressure bifurcation tunnel with reinforced concrete lining, design for machine-bored tunnel, design for special rock mass and poor geological tunnel, plugging body design, grouting and seepage control and drainage, safety monitoring, and operation and maintenance.

The main technical contents revised are as follows:

—Adding relevant design contents for high-pressure tunnel with reinforced concrete lining, machine-bored tunnel, and special rock mass and poor geological tunnel.

—Adding relevant technical requirements for filling and emptying of power tunnel.

—Adding relevant contents regarding calculations of discharge capacity, aeration water depth, external water pressure, and stresses of circular tunnel lining under uniform external water pressure.

—Adding design situations such as the construction and maintenance periods of tunnel, and operation period of water release tunnel, and the action and partial factors of backfill grouting.

—Adding seepage stability requirements for plugging body design.

—Revising relevant contents of the layout and hydraulic design for high flow velocity tunnel.

—Revising the stability calculation formula of surrounding rock block, and adjusting the partial factor formula to safety factor formula.

—Deleting Appendix A "Engineering Geological Classification of Surrounding Rock", Appendix B "Materials", Appendix F "Type and Parameter of Anchor-shotcrete Support", and Appendix I "Lining Calculation for Circular

Free-Flow Tunnel and Non-circular Tunnel" in the previous version; and merging the provisions of Appendix D and Appendix E in the previous version into the text of this code.

—Integrating Appendix J "Concrete Lining Cracking and Its Prevention Measures" in the previous version into the relevant articles and items of this code.

National Energy Administration of the People's Republic of China is in charge of the administration of this code. China Renewable Energy Engineering Institute has proposed this code and is responsible for its routine management. Energy Sector Standardization Technical Committee on Hydropower Investigation and Design is responsible for the explanation of specific technical contents. Comments and suggestions in the implementation of this code should be addressed to:

China Renewable Energy Engineering Institute
No. 2 Beixiaojie, Liupukang, Xicheng District, Beijing 100120, China

Chief development organization:

POWERCHINA Chengdu Engineering Corporation Limited

Participating development organizations:

POWERCHINA Huadong Engineering Corporation Limited

POWERCHINA Northwest Engineering Corporation Limited

POWERCHINA Zhongnan Engineering Corporation Limited

Chief drafting staff:

HUANG Yankun	YANG Huaide	HAO Yuanlin	LIU Yue
YOU Xiang	XIE Jinyuan	LIU Yuan	ZHANG Yang
LIU Changgui	LIU Lijuan	DU Zhenyu	CHEN Xiangrong
YI Bo	ZHAO Lu	LI Yun	CHEN Xugao
JU Lin			

Review panel members:

DANG Lincai	HAO Jungang	ZENG Xionghui	LIU Shanjun
WANG Jianhua	WANG Kangzhu	TIAN Zhenghai	NING Huawan
LI Linian	CHEN Lifen	WANG Ke	XIONG Chungeng
LI Yun	YANG Xiaolong	CHEN Wenhua	DU Xiaokai
ZHENG Linping	HE Shuangxi		

Contents

1 General Provisions

1.0.1 This code is formulated with a view to standardizing the design of hydraulic tunnel, to achieve safety, reliability, technological advancement and cost-effectiveness.

1.0.2 This code is applicable to the design of the construction, renovation and extension of hydraulic tunnels for hydropower projects.

1.0.3 The grade of hydraulic tunnel shall be in accordance with the current standards of China GB 50201, *Standard for Flood Control*; and DL 5180, *Classification & Design Safety Standard of Hydropower Projects.*

1.0.4 The hydraulic tunnel design shall make full use of self-stability, bearing capacity and impermeability of surrounding rock.

1.0.5 The structural design of pressure hydraulic tunnel with steel lining shall comply with the current sector standard NB/T 35056, *Design Code for Steel Penstocks of Hydroelectric Stations*.

1.0.6 The design of hydraulic tunnel shall meet the requirements of overall planning and environmental protection of the project.

1.0.7 In addition to this code, the design of hydraulic tunnel shall comply with other current relevant standards of China.

2 Terms

2.0.1 hydraulic tunnel

enclosed water passage excavated in mountains or underground in a hydropower and water conservancy project for water conveyance, power generation, irrigation, flood discharge, diversion, reservoir emptying, sediment flushing, etc.

2.0.2 pressure tunnel

hydraulic tunnel fully filled with water, whose wall bears the water pressure

2.0.3 free-flow tunnel

hydraulic tunnel partially filled with water

2.0.4 high-pressure tunnel

tunnel with a head no less than 100 m

2.0.5 high flow velocity tunnel

tunnel with a flow velocity greater than 20 m/s

2.0.6 tunnel support

engineering measure to reinforce the tunnel surrounding rock by structures or elements and other materials

2.0.7 anchor-shotcrete support

engineering measure to reinforce the rock mass by anchor bolts and shotcrete

2.0.8 primary support

support installed immediately after the excavation of a tunnel

2.0.9 secondary support

support after primary support according to the monitoring results or tunnel function

2.0.10 random anchor bolt

anchor bolt installed in partial surrounding rock to prevent rock collapse or sliding

2.0.11 systematic anchor bolt

anchor bolt installed regularly at a certain spacing on the excavation face according to the stability requirements of the surrounding rock

2.0.12 preset anchor bolt

anchor bolt pre-installed in the surrounding rock around the heading face for stability of the next advance

2.0.13 lining

engineering measure for support in underground works to consolidate surrounding rock or smooth flow surface using concrete, reinforced concrete, etc.

2.0.14 backfill grouting

engineering measure using grout to fill the gaps and voids between concrete lining and surrounding rock, to improve the compactness of surrounding rock or structure

2.0.15 joint grouting

engineering measure injecting grout into the joint between concrete blocks by embedding pipelines or other ways, to enhance the integrity of the structure and improve the conditions of force transmission

2.0.16 contact grouting

engineering measure injecting grout into the gaps between concrete and bedrock by embedding pipelines or other ways, to enhance the bonding of contact surfaces

2.0.17 consolidation grouting

engineering measure using grout to reinforce the surrounding rock with geological defects such as open joints or fracture zones, to improve rock integrity and bearing capacity, and to reduce seepage

2.0.18 hydraulic fracturing

physical phenomenon that pressure water in a hydraulic tunnel results in expansion, spreading and interpenetration of the existing joints and fractures in the rock mass

2.0.19 machine-bored tunnel

tunnel excavated by tunnel boring machine (TBM). TBM is an advanced construction machine using rotary cutters to excavate and break the rock or the overburden in the tunnel and form a full-sized tunneling section

2.0.20 cavitation

process of the formation, development and collapse of vapor or bubble in the liquid or at the liquid-solid interface when the local pressure is reduced to the saturated vapor pressure of the liquid

2.0.21 cavitation erosion

erosion damage of the solid boundary caused by cavitation

2.0.22 rock burst

explosive breaking and popping of rock caused by sudden and violent release of elastic strain energy accumulated in rock mass in certain conditions

3 Basic Requirements

3.0.1 The design of hydraulic tunnels shall collect the data on hydrology, sediment, hydropower energy economy, topography, geology, earthquake, ecological environment, soil and water conservation, construction conditions, hydraulic steel structure, electromechanical equipment, construction materials, operation modes, etc., according to the project layout, tunnel function and requirements at different design stages.

3.0.2 The geological investigation for the inlet, outlet and tunnel alignment shall comply with the current national standard GB 50287, *Code for Hydropower Engineering Geological Investigation* according to the complexity of topography and geology, grade of tunnel, and different design stages. The in situ stress test, hydraulic fracturing test or other tests shall be conducted at the selected representative locations on site for Grade 1 high-pressure hydraulic tunnels and high-pressure concrete bifurcation tunnels.

3.0.3 The following basic geological data on hydraulic tunnel shall be collected at the preliminary design stage:

1 Surrounding rock characteristics, geological structures and in situ stress along the tunnel alignment.

2 Karst and hydrogeology along the tunnel alignment.

3 Stability of the tunnel inlet, outlet and relevant slope.

4 Geological phenomena that impact the tunnel safety, such as the karst cave, water inrush, rock burst, high ground temperature, harmful gas, and radioactive substances.

3.0.4 During the excavation of hydraulic tunnel, the detailed geological logging shall be conducted in time according to the actual conditions, and the geological data shall be collected and checked to forecast the geological conditions for construction. For the tunnel section with complex geological conditions, the pilot tunnel, advance borehole, advance geophysical exploration, etc. should be used in construction to ascertain the geological conditions.

3.0.5 The classification of surrounding rock masses in hydraulic tunnel shall comply with the current national standard GB 50287, *Code for Hydropower Engineering Geological Investigation.*

3.0.6 The parameters of shotcrete, anchor bolt, concrete and reinforcement, etc., used in the hydraulic tunnel design shall be in accordance with the current standards of China GB 50086, *Technical Code for Engineering of Ground Anchorages and Shotcrete Support*; and DL/T 5057, *Design Specification for Hydraulic Concrete Structures.*

3.0.7 The rock trap should be set in a long water conveyance and power generation tunnel.

4 Tunnel Layout

4.1 Tunnel Line Selection

4.1.1 The tunnel alignment shall be determined through techno-economic comparison according to the purpose and characteristics of the tunnels, considering the topography, geology, rock cover, ecological environment, soil and water conservation, project general layout, hydraulics, construction, operation, structures along the alignment, etc.

4.1.2 Under the premise of satisfying the requirements of the project general layout, the tunnel alignment should be arranged in a region with simple geological structure, stable and intact rock mass and favorable hydrogeological conditions, and being convenient for construction and access, and shall also meet the following requirements:

1 There should be a relative large included angle between the tunnel alignment and the strike of bedding planes, major discontinuities and weak zones, which should not be less than 30°, and for thinly bedding planes with poor cementation and steeply-dipping angle, the included angle should not be less than 45°. If the included angle is less than the specified values, engineering measures shall be taken.

2 The effects of in situ stress on the stability of surrounding rock shall be considered for tunnels located in high stress zone. The tunnel alignment should be parallel to the direction of maximum horizontal in situ stress, or meet it at a small included angle.

3 The included angle between the outlet axis of water release tunnel, such as flood discharge tunnel, emptying tunnel, and sediment flushing tunnel, with the midstream of downstream river reach shall be determined according to the topography, geology and hydraulics at the tunnel outlet.

4.1.3 For shallow cover tunnels, the open channel, conduit or deepening cover tunnel may be adopted. The final scheme should be selected through techno-economic comparison.

4.1.4 The minimum rock cover of tunnel shall be analyzed and determined according to the topography, geology, uplift resistance and impermeability of the rock mass, hydrostatic pressure in tunnel, and support pattern, and shall meet the following requirements:

1 The minimum rock cover (Figure 4.1.4) of the pressure tunnel shall be determined as per the hydrostatic pressure less than the weight of rock mass over the tunnel crown, which should be calculated by the following formula:

$$C_{RM} = \frac{h_s \gamma_w F}{\gamma_R \cos \alpha} \tag{4.1.4}$$

where

C_{RM} is the minimum rock cover excluding the depth of completely and highly weathered rock mass (m);

h_s is the hydrostatic head in tunnel (m);

γ_w is the unit weight of water (N/mm^3);

γ_R is the unit weight of rock mass (N/mm^3);

F is the empirical coefficient, which should be taken as 1.30 to 1.50, and the greater value should be taken where geological conditions are poor;

α is the dip angle of the river bank slope (°), which is taken as 60° if α>60°.

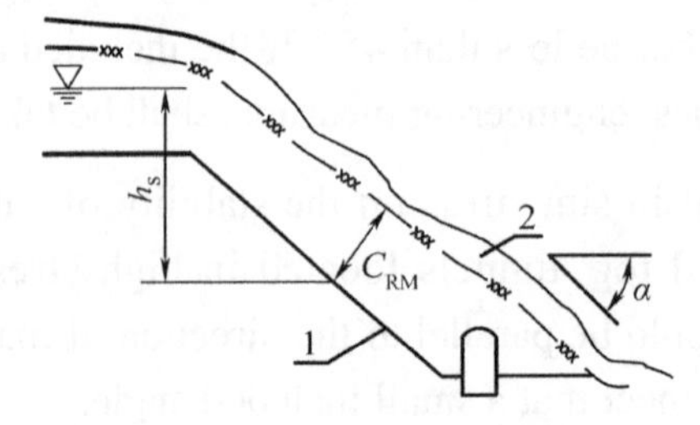

Key

1 pressure tunnel

2 completely and highly weathered rock mass

Figure 4.1.4 Minimum rock cover of a pressure tunnel

2 The minimum rock cover of the pressure tunnel shall ensure that the surrounding rock does not result in seepage instability, and the seepage gradient of the surrounding rock shall meet the seepage stability requirements. For the tunnels of important projects or with particularly high water head, special study should be conducted on the seepage stability of surrounding rock under high water pressure based on the high-pressure seepage test.

3 The high-pressure tunnel and bifurcation tunnel shall also meet the requirement that the hydrostatic pressure is less than the minimum in situ stress of the surrounding rock.

4 For the pressure tunnels that fail to satisfy Items 1 to 3 of this article, reasonable construction procedures and engineering measures shall be taken to ensure the safety in construction and operation periods.

5 Under the premise of ensuring the safety in construction and operation periods, the minimum rock cover of the free-flow tunnel need not be specified.

4.1.5 Where the discharge is relatively large and the geological conditions are unfavorable for large cross-section tunnel excavation, two- or multi-tunnel layout may be adopted, which shall meet the following requirements:

1 The layout scheme shall be determined by comprehensive analysis according to hydraulic and rock mass stress conditions, arrangement of head and tail structures, construction and operation conditions, possibility of staged operation, project cost, construction period, etc.

2 The parallel layout scheme for multiple flood discharge tunnels shall be determined by comprehensive analysis according to flow conditions at inlets of adjacent tunnels and excavation coordination, arrangement of structures at outlet, anti-scouring capability of downstream riverbed, etc., which shall be coordinated with the auxiliary structures.

4.1.6 The thickness of rock mass between adjacent tunnels or spatial crossing tunnels shall be determined by comprehensive analysis according to the layout, topography and geology, stress and deformation of surrounding rock, hydraulic fracturing and seepage stability of rock mass, section shape and dimensions of tunnels, construction methods, operation and maintenance conditions, etc. The thickness of rock mass should not be less than 2 times the excavation diameter (or span) of the smaller tunnel, and shall ensure that no seepage instability or hydraulic fracturing occurs in the operation period.

4.1.7 Tunnels passing through dam foundation, abutments or the foundation of other structures shall meet the following requirements:

1 The thickness of rock mass between the tunnel and other structures shall meet the structural and seepage control requirements.

2 For a pressure tunnel, the layout relationship between the tunnel and grout curtain of dam or powerhouse shall be analyzed, and the seepage control and drainage measures shall be determined according to the surrounding rock conditions and adjacent structures.

4.1.8 A steel lining section shall be arranged at the high-pressure tunnel upstream of underground powerhouse, and its length should not be less than 0.25

times the maximum static water head.

4.1.9 The scheme of tunnel alignment across the gullies shall be determined through techno- economic comparison between alternatives of gully bypassing and crossing according to the topography, geology, hydrology, construction and seismic conditions. The gully-crossing scheme, if adopted, shall consider the effects of flood and debris flow in the gully.

4.1.10 The horizontal bend section of hydraulic tunnel shall meet the following requirements:

1 For a low velocity free-flow tunnel, the radius of bend should not be less than 5 times the tunnel diameter (or span), and the angle of bend should not be greater than 60°. The straight sections shall be provided at the start and end of the bend, and the section length should not be less than 5 times the tunnel diameter (or span). For a low velocity pressure tunnel, the bend radius may be appropriately reduced, but should not be less than 3 times the tunnel diameter (or span).

2 For a high velocity free-flow tunnel, the tunnel alignment should be arranged as a straight line.

3 For a high velocity pressure tunnel, the parameters of bend should be determined by hydraulic model tests.

4 For a tunnel excavated by the TBM or with mucking rail transport, the radius and angle of bend shall meet the requirements for the TBM and rail transport.

4.1.11 For a high velocity tunnel, the type of the vertical curve and radius of bend should be determined by hydraulic model tests. For a low velocity free-flow tunnel, the radius of bend should not be less than 5 times the tunnel diameter (or span). For a low velocity pressure tunnel, the radius of bend should not be less than 3 times the tunnel diameter (or span). The connection arrangement of vertical curves shall consider hydraulic conditions, construction methods, etc.

4.1.12 The tunnel longitudinal slope shall be determined according to the operation, hydraulic conditions, upstream and downstream transition, construction and maintenance conditions, etc., and the adverse slope should be avoided.

4.1.13 The adits shall be arranged in the long tunnel constructed by drilling and blasting method. The number and lengths of adits shall be determined through techno-economic analysis according to the topography and geology along the alignment, accessibility, construction conditions, bill of quantities, construction schedule, etc.

4.2 Tunnel Inlet and Outlet Layout

4.2.1 The layout of tunnel inlets and outlets should be determined according to the requirements for function, project general layout, topography and geology, hydraulics, deposition and scouring control, trash control, etc.

4.2.2 Tunnel portals should be arranged on the bank slopes with simple geological structure, shallow overburdens, shallow rock weathering and unloading, and should avoid such unfavorable geological zones as faults, landslide, collapse and debris flow.

4.2.3 Engineering measures such as reinforcement, water control and drainage shall be taken for the slopes at the tunnel inlet and outlet as required by the results of slope stability analysis, and high slope excavation should be avoided.

4.2.4 The inlet layout of the headrace tunnel shall comply with the current sector standard DL/T 5398, *Design Specification for Intake of Hydropower Station*. The inlet layout of other hydraulic tunnels may be open or pressure type, the inlet layout shall meet the following requirements:

1 The flow boundary of open inlet shall be smooth. The design of control weir (sluice) shape shall comply with the current sector standard DL/T 5166, *Design Specification for River-Bank Spillway*.

2 The short pressure inlet should include inlet bell mouth section, gate slot section and downward slope section. The contraction type shall be adopted in the downward slope section upstream of the service gate, where the pressure distribution shall decrease gradually along the alignment to avoid cavitation.

3 The long pressure inlet should adopt the elliptic curve type with three-direction contraction on the crown and both sidewalls.

4 The unfavorable vortex and backflow in front of the inlet shall be avoided. The inlet with relative complex topography, layout or shape shall be verified by the hydraulic model test.

4.2.5 The layout of the tunnel portal for a pumped storage power station shall meet the requirements for two-way flow, and should be determined by the hydraulic model test.

4.2.6 The outlet of pressure water release tunnel shall be well transitioned with downstream channel, and the shape of the tunnel outlet section shall meet the following requirements:

1 The configuration of tunnel outlet section should be determined according to the hydraulic conditions, type and layout of the service gate, and its opening mode.

2 The outlet cross-sectional area should be 85 % to 90 % of that of the tunnel. If the geometry varies greatly along the tunnel and the flow condition in tunnel is poor, the contraction percentage should be 80 % to 85 %. For important hydraulic tunnels, hydraulic model tests shall be performed.

4.2.7 The design of air supply and exhaust facility of hydraulic tunnels shall meet the following requirements:

1 For a pressure tunnel, the air vent shall be provided at the inlet or outlet. The cross-sectional area of the air vent may be calculated by the formulae in the current sector standard DL/T 5398, *Design Specification for Intake of Hydropower Station*.

2 For a high velocity free-flow tunnel with aerators, dedicated air supply tunnels leading to the atmosphere should be provided at different sections according to flow velocity distribution. Locations and number of air supply tunnels should be determined by the hydraulic model tests. The dedicated air supply tunnel should be provided for the underground service gate chamber. The average wind speed of air supply should be less than 40 m/s, and the maximum should be less than 60 m/s.

4.2.8 The outlet layout of water release tunnel shall ensure that the released water does not impact the safety and normal operation of other structures and facilities.

4.3 Multipurpose Tunnel

4.3.1 The tunnel layout scheme shall study the feasibility and economy of combining the temporary and permanent functions and using one tunnel for multiple purposes according to the tunnel purpose, operation and construction conditions.

4.3.2 The design of tunnel combining temporary and permanent functions shall meet the requirements for temporary discharging and permanent operation.

4.3.3 The design of diversion tunnel should study the possibility of full or partial section to be used as the permanent hydraulic tunnel such as flood discharge tunnel, emptying tunnel, sediment flushing tunnel and tailrace tunnel.

4.3.4 The flood discharge tunnel to be reconstructed may adopt the internal energy dissipators such as vortex flow, thick orifice, and orifice ring which shall be determined according to the scale of flood discharge, water head and seepage control layout at both banks, topography and geology of riverbed and both banks, water depth downstream, and water level amplitude, and shall be verified by the hydraulic model test.

5 Shape and Size of Cross Section

5.1 General Requirements

5.1.1 The shape and size of a tunnel cross section shall be determined by techno-economic analysis according to the tunnel purposes, hydraulic conditions, engineering geology and hydrogeology, in situ stress, reinforcing forms of surrounding rock, construction methods, etc.

5.1.2 The flow pattern in a tunnel shall meet the following requirements:

1 For a pressure tunnel, the flow pattern alternation between free-flow and pressure-flow shall not be allowed, and the minimum pressure head at the crown along the full tunnel shall not be less than 2.0 m under the most unfavorable operating condition.

2 For a high flow velocity water release tunnel, the flow pattern alternation between free-flow and pressure-flow shall not be allowed. For a low flow velocity water release tunnel, such flow pattern may be allowed under the check flood condition.

3 For a tailrace tunnel and a diversion tunnel, the flow pattern alternation between free-flow and pressure-flow may be allowed after study and demonstration.

5.1.3 For tunnels with alternation between free-flow and pressure flow, the structural strength, stiffness and integrity shall be enhanced in design.

5.2 Cross Section Shape

5.2.1 Circular section should be adopted for pressure tunnels. If the diameter of tunnel is small, and internal and external water pressure is not high, other section shapes convenient for construction may be adopted. The inverted U-shape section should be adopted for free-flow tunnels. The crown angle should be 90° to 180°. If the tunnel is in poor geological conditions or the included angle between the tunnel axis and the direction of rock strata is less than the value specified in Article 4.1.2 of this code, circular or horseshoe-shape section should be selected.

5.2.2 The height-width ratio of the tunnel section, which may be selected according to geology, in situ stress and hydraulic conditions, should be taken as 1.0 to 1.5 and meet the following requirements:

1 A lower ratio should be adopted if the horizontal in situ stress is greater than the vertical one.

2 A higher ratio should be adopted if the vertical in situ stress is greater than the horizontal one or there are thinly bedding planes with poor cementation and gently-dipping angle.

3 A higher ratio should be adopted for tunnels with free-flow inlet and outlet and great variation in water level.

5.2.3 Cross-section shapes may vary for a relatively long tunnel, but the shapes should not change frequently, and the connections between different sections shall meet the following requirements:

1 A transition section shall be provided between two different sections and its boundary shall be a gentle curve to facilitate construction.

2 The expansion or contraction angle of the pressure tunnel transition section should be 6° to 10°. The length should not be less than 1.5 times the tunnel diameter (or span).

3 The transition section shape of the high-velocity free-flow tunnel shall be determined by hydraulic model tests.

5.3 Cross Section Size

5.3.1 The cross-sectional size of the water conveyance tunnel in a hydropower station shall be determined by energy economy comparisons.

5.3.2 The cross-sectional size of the water release tunnel shall satisfy the tunnel discharge capacity under various operating conditions.

5.3.3 The cross-sectional size of the diversion tunnel shall be determined by techno-economic comparison according to diversion discharge, invert elevation of inlet and outlet, cofferdam height, hydraulic transition of outlet, etc.

5.3.4 The minimum cross-sectional size of tunnels shall be determined by taking into consideration the cross-sectional shape and construction methods. The diameter of circular section tunnel should not be less than 2.0 m, and the non-circular section tunnel should be no less than 2.0 m in height and no less than 1.8 m in span.

5.3.5 For low velocity free-flow tunnels with good ventilation condition, the cross-sectional size shall meet the following requirements:

1 For steady flow, the area above the water surface profile should not be less than 15 % of the tunnel section area, and the height above water surface profile shall not be less than 0.4 m.

2 For unsteady flow, the area and height above the water surface profile may be decreased properly if the surge wave has already been taken into account in water surface profile calculation.

3 For tunnels unlined, with anchor-shotcrete support or longer than 1,000 m, the area and height above the water surface profile may be increased properly.

5.3.6 Aeration effects shall be considered in determining the cross-sectional size of high-velocity free-flow tunnels. The area above the aerated water surface profile should be 15 % to 25 % of the total cross-sectional area, and may be determined based on the analysis of Froude number. The specific cross-section size should be determined by hydraulic model tests. For an inverted U-shape section, the wave peak shall be limited to the height of the vertical sidewall when there are shock waves in the flow.

6 Hydraulic Design

6.1 Principles for Hydraulic Calculation

6.1.1 Hydraulic calculation items of hydraulic tunnels shall include head loss, hydraulic transient process, discharge capacity, connection of upstream and downstream water surfaces, energy grade line, water surface profile, aeration, water filling and emptying, etc. Calculation items shall be selected based on the purposes and design stage of the tunnel.

6.1.2 Head loss calculation of hydraulic tunnels shall comply with Appendix A of this code. For a tunnel of complicated shape, an inlet and outlet of a pumped-storage power station, and a layered water intake, the head loss should be determined by hydraulic model tests.

6.1.3 The discharge capacity calculation for the hydraulic tunnel with an open inlet shall comply with the current sector standard DL/T 5166, *Design Specification for River-Bank Spillway.* The calculation of discharge capacity for hydraulic tunnel with pressure inlet shall comply with Appendix B of this code. The discharge capacity of the free-flow tunnel without an inlet control weir (sluice) may be calculated as per the open channel flow.

6.1.4 The step method or other methods may be used for the calculation of water surface profile of free-flow tunnels. The aeration water depth calculation for a high-velocity free-flow tunnel may be conducted in accordance with Appendix C and determined through comprehensive analysis based on unit discharge, Froude number, bottom slope, air-entraining and aeration conditions, engineering practices, etc.

6.1.5 For hydraulic tunnels with high flow velocity, large discharge and complicated flow conditions, the hydraulic design shall be verified by hydraulic model tests. Depressurized model tests should be carried out in the case of prominent cavitation or cavitation erosion.

6.2 Design of Cavitation Erosion Control for Water Boundary with High Flow

6.2.1 The flow cavitation number σ may be calculated by the following formulae:

$$\sigma = \frac{h_0 + h_a - h_v}{\frac{v_0^2}{2g}} \tag{6.2.1-1}$$

$$h_a = 10.33 - \frac{\nabla}{900} \tag{6.2.1-2}$$

where

h_0 is the pressure water column at the calculation section (m);

h_a is the atmospheric pressure water column at the calculation section (m);

h_v is the vapor pressure water column at the corresponding water temperature (m), which may be taken as per Table 6.2.1;

$\frac{v_0^2}{2g}$ is the average velocity head at the calculation section (m);

∇ is the elevation of the calculation section (m).

Table 6.2.1 Relationship between water temperature and vapor pressure water column

Water temperature (°C)	**0**	**5**	**10**	**15**	**20**	**25**	**30**	**40**
h_v **(m)**	0.06	0.09	0.13	0.17	0.24	0.32	0.43	0.75

6.2.2 For each part of a high flow velocity hydraulic tunnel, the flow cavitation number σ should be greater than the incipient cavitation number σ_i, and shall meet the following requirements:

1 σ_i may be determined by depressurized model tests and engineering practices.

2 The cross-sectional shape for the position where σ is less than σ_i shall be further optimized, or other engineering measures shall be taken to mitigate cavitation erosion.

6.2.3 In the hydraulic design of a high flow velocity tunnel, attention shall be paid to the following positions or sections prone to cavitation erosion:

1 The inlet, gate slot, transition section, bifurcation section, bend section, positions with abrupt changes in flow boundary and with abrupt expansion or fall at the outlet of a pressure tunnel.

2 The pier, gate slot, vertical curve section connecting with steep slopes, invert and sidewall downstream of reverse curve section, diffusion section, slope changing section or positions with abrupt changes in flow boundary, connection section of a free-flow tunnel.

3 The high flow velocity tunnel section with flow cavitation number smaller than 0.30 or flow velocity greater than 30 m/s.

4 Energy dissipaters at the outlet.

6.2.4 Cavitation erosion prevention design shall meet the following requirements:

1 The shape with a smaller flow incipient cavitation number σ_i should be selected, and the high flow velocity tunnel section should be shortened.

2 The shape, structural materials, possible continuous work duration, etc. shall be considered for surface irregularity control criteria. The local surface irregularities may be controlled as per Table 6.2.4.

3 Aerators should be provided at the positions prone to cavitation erosion. Suitable aerators shall be set at the high flow velocity tunnel sections with flow cavitation number less than 0.30 or flow velocity greater than 35 m/s. The layout of aerators shall meet the following requirements:

1) Multiple aerators may be set for the high flow velocity tunnel with a long chute section. The protective length of an aerator may be 70 m to 100 m for the reverse curve section, and 100 m to 200 m for the straight section.

2) The aeration concentration shall be at least 3 % for the protection area covered by the aerator, and at least 5 % for the positions with extra high requirements. There shall be a desirable bubble distribution in the aerated flow. For tunnel sections with a flow velocity greater than 40 m/s, attention shall be paid to the aeration concentration of sidewall.

3) Undesirable flow pattern caused by aerators shall be avoided.

4) The aeration cavity shall keep stable. The ventilation well (slot) and aeration step (slot) shall be in smooth ventilation, without blocking.

5) The designed ventilation well (slot) should be simple and safe, the average wind speed of the ventilation system should be less than 40 m/s and the maximum wind speed should be less than 60 m/s.

6) The location, type, and size of aerators should be determined by model tests, and model scaling effects should be considered.

4 High strength concrete, fiber concrete, epoxy mortar, steel plate, etc. should be used for cavitation erosion prevention. Cracks in the flow surface of concrete shall be avoided. For high flow velocity tunnels in sediment-laden river areas, the materials with proper cavitation erosion and abrasion resistance shall be adopted according to the combined effects of sediment-laden flow abrasion, bedload jumping impact and cavitation erosion.

5 Unfavorable working conditions shall be avoided in operation.

Table 6.2.4 Local surface irregularities

Flow cavitation number		$\sigma > 0.60$	$0.35 \le \sigma \le 0.60$	$0.30 \le \sigma < 0.35$	$0.20 \le \sigma < 0.30$		$0.15 \le \sigma < 0.20$		$0.10 \le \sigma < 0.15$	$\sigma < 0.10$	
Aerator		–	–	–	Not required	Required	Not required	Required	Not required	Required	Revise the design or aerate adequately
Offset height (mm)		≤ 25	≤ 12	≤ 8	< 6	< 15	< 3	< 10	Revise the design	< 6	
Treated slope	Upstream slope	–	1/10	1/30	1/40	1/8	1/50	1/10		1/10	
	Downstream slope	–	1/5	1/10	1/10	1/4	1/20	1/5		1/8	
	Side slope	–	1/2	1/3	1/5	1/3	1/10	1/3		1/4	

NOTE Offset height should be checked with a 2-m guiding rule.

6.2.5 In addition to Articles 6.2.1 to 6.2.4 of this code, cavitation erosion prevention design of high flow velocity tunnels shall comply with the current sector standard DL/T 5207, *Technical Specification for Abrasion and Cavitation Resistance of Concrete in Hydraulic Structures*.

7 Unlined Tunnel and Anchor-Shotcrete Tunnel

7.1 General Requirements

7.1.1 Unlined and anchor-shotcrete tunnel shall meet the stability requirements for surrounding rock and seepage. Water exfiltration from the tunnel shall not endanger the adjacent slope or structures, or cause any damage to the environment.

7.1.2 Anchor-shotcrete support should not be used as permanent support for a tunnel in any of the following cases:

1 The tunnel section with long-term large-area water gushing.

2 The tunnel section with its shotcrete layer prone to erosion or in swelling strata.

3 The tunnel section of great importance or with complex geological conditions.

4 The tunnel section with high in situ stress.

5 The tunnel section with special requirements.

7.1.3 For Class Ⅰ or Class Ⅱ surrounding rock, the tunnel with a diameter or span less than 5 m need not be supported, the tunnel with a diameter of 5 m to 10 m should be supported with shotcrete, and the tunnel with a diameter larger than 10 m should be supported with anchor-shotcrete. Random anchor bolts or bolt bundles shall be used for reinforcement in the case of local unstable block.

7.1.4 For Class Ⅲ surrounding rock, the tunnel should be supported by anchor-shotcrete with steel mesh. For Class Ⅳ surrounding rock, the tunnel should be supported by reinforced concrete lining, and anchor-shotcrete, steel mesh or steel rib may also be used in a combined manner after analysis according to the scale of the tunnel and geological conditions. For Class Ⅴ surrounding rock, the tunnel shall be supported by reinforced concrete lining or other permanent lining.

7.1.5 The cross-sectional size of unlined and anchor-shotcrete tunnel should be determined on the principle of head loss being equal to that of concrete lined section.

7.1.6 The design of the tunnel with no support or with anchor-shotcrete support should be performed by engineering analogy. For Grade 1 tunnel or tunnel with a diameter (or span) greater than 10 m, theoretical calculation, numerical calculation and monitoring measurement should be carried out.

7.1.7 The overall stability of surrounding rock should be calculated by the numerical analysis method. The stability of the rock mass which might be locally unstable may be calculated by the block limit equilibrium method.

7.1.8 In preliminary design, the type and parameters of anchor-shotcrete support may be selected in accordance with the current national standard GB 50086, *Technical Code for Engineering of Ground Anchorages and Shotcrete Support*, taking into account engineering geological conditions, tunnel size, and tunnel purpose. In detail design, they shall be adjusted according to the geological conditions revealed on site.

7.1.9 The tunnel section with no support or with anchor-shotcrete support near the portal shall be strengthened by reinforced concrete lining or other measures. The reinforced section length should be greater than the depths of unloading zone and highly weathered zone in the portal section, and should be greater than the tunnel diameter (or span).

7.1.10 On the invert of the tunnel unlined or supported by anchor-shotcrete, a concrete slab with a thickness of no less than 0.2 m should be placed.

7.1.11 A rock trap shall be provided at the end of the unlined and anchor-shotcrete headrace tunnel, and its volume may be determined considering the unlined section length, surrounding rock conditions, hydraulic conditions, debris clearing frequency and convenience, etc.

7.1.12 The hydraulic design of rock trap should meet the following requirements:

1 The flow disturbance on the tunnel cross section is small.

2 The flow disturbance within the rock trap is small.

3 A baffle plate should be set in the rock trap.

4 For an important project, hydraulic model test should be performed for the rock trap.

7.1.13 The anchor-shotcrete support should be applied shortly after excavation, and safety monitoring shall be provided. The irregularity of shotcrete surface should be less than 0.15 m.

7.2 Shotcrete Support

7.2.1 The design strength of shotcrete shall not be inferior to C20. For the support of large tunnel or under special conditions, the shotcrete should not be inferior to C25. The 1 day compressive strength of shotcrete shall not be less than 8 N/mm^2. The bonding strength between the shotcrete and surrounding rock should not be less than 0.8 N/mm^2 for structural type and 0.2 N/mm^2 for protective type, respectively.

7.2.2 The shotcrete thickness should be 0.10 m to 0.15 m. The minimum thickness shall be 0.05 m, and the maximum thickness should be 0.20 m.

7.2.3 The flow velocity in a permanent tunnel supported by shotcrete should not be greater than 8 m/s. For a temporary tunnel, the flow velocity should not be greater than 12 m/s.

7.2.4 The tunnel section with special geological conditions, such as the surrounding rock prone to plastic deformation or located in high in situ stress zone, should be supported by steel fiber shotcrete, synthetic fiber concrete or other special materials. The selection of special material shall comply with the current national standard GB 50086, *Technical Code for Engineering of Ground Anchorages and Shotcrete Support*, and the composition and mix proportion of the materials should be determined by site tests.

7.2.5 The surface of steel fiber shotcrete should be covered with a layer of plain shotcrete, whose strength shall not be inferior to that of the steel fiber shotcrete and thickness should not be less than 0.03 m.

7.3 Anchor Bolt Support

7.3.1 Anchor bolts should be applied to reinforce the surrounding rock which is hard and intact but locally unstable. Anchor bolt bundle or anchor cable may be applied to reinforce the surrounding rock with a large and deep loosening range. Local weak rock mass, such as fault and dense joints, may be reinforced by combining anchor bolt with anchor bolt bundle or anchor cable, and may be provided with steel mesh.

7.3.2 When the surrounding rock is reinforced by anchor bolts, anchor bolt bundle or anchor cable, the stability safety factor of surrounding rock shall not be less than the values shown in Table 7.3.2. The stability safety factor may be calculated by the following formulae:

1 When the collapse-type unstable block is reinforced by anchor bolts, anchor bolt bundle, or anchor cable, the stability safety factor may be calculated as follows:

1) For cement mortar anchor bolts or anchor bolt bundle, the stability safety factor of the block may be calculated by the following formula:

$$K = \frac{nA_x f_y \cos\alpha}{G} \tag{7.3.2-1}$$

where

K is the stability safety factor of the block;

G is the dead weight of the unstable block (N);

n is the number of anchor bolts or anchor bolt bundles;

A_x is the sectional area of single anchor bolt with cement mortar or single anchor bolt bundle (mm^2);

f_y is the design value of tensile strength for anchor bolt with cement mortar (MPa);

α is the included angle between the anchor bolt or anchor bolt bundle and the vertical direction (°).

2) For prestressed anchor bolts or anchor cables, the stability safety factor of the block may be calculated by the following formula:

$$K = \frac{nA_y\sigma_{con}\cos\alpha}{G} \quad (7.3.2\text{-}2)$$

where

n is the number of prestressed anchor bolts or anchor cables;

A_y is the sectional area of the single prestressed anchor bolt or the single prestressed anchor cable (mm^2);

σ_{con} is the control value of tensile stress for the prestressed anchor bolts or the prestressed anchor cables (MPa), which shall be taken in accordance with the current sector standard DL/T 5057, *Design Specifications for Hydraulic Concrete Structures*;

α is the included angle between the prestressed anchor bolt or prestressed anchor cable and the vertical direction (°).

2 When the sliding-type unstable block is reinforced by anchor bolts, anchor bolt bundle, or anchor cables, the stability safety factor may be calculated as follows:

1) For anchor bolts with cement mortar or anchor bolt bundle, the stability safety factor may be calculated by the following formula:

$$K = \frac{fG_2 + nA_s f_{gv} + CA}{G_1} \quad (7.3.2\text{-}3)$$

where

G_1 is the component force of the unstable block dead weight acting parallel to the sliding plane (N);

G_2 is the component force of the unstable block dead weight acting perpendicular to the sliding plane (N);

A_s is the sectional area of the single anchor bolt or the single anchor bolt bundle (mm^2);

A is the area of the sliding plane of the block (mm^2);

f is the friction coefficient of the sliding plane;

C is the cohesion strength of the block on the sliding plane (MPa);

f_{gv} is the design value of shear strength for the anchor bolt (Mpa), which may be taken as 0.5 times the design value of tensile strength of the anchor bolt.

2) For prestressed anchor bolts or anchor cables, the stability safety factor of the block may be calculated by the following formula:

$$K = \frac{f(G_2 + P_n) + P_t + CA}{G_1} \quad (7.3.2\text{-}4)$$

where

P_t is the component force of the total pressure acting on the unstable block loaded by the prestressed anchor bolts or anchor cables acting along the sliding direction (N);

P_n is the component force of the total pressure acting on the unstable block loaded by the prestressed anchor bolts or anchor cables acting perpendicular to the sliding direction (N).

Table 7.3.2 Safety factor of surrounding rock block stability

Structural safety class	Collapse-type block			Slipping-type block		
	Persistent design situation	Transient design situation	Accidental design situation	Persistent design situation	Transient design situation	Accidental design situation
I	2.00	1.90	1.70	1.80	1.65	1.50
II	1.90	1.70	1.60	1.65	1.50	1.40
III	1.70	1.60	1.50	1.50	1.35	1.25

7.3.3 The anchor bolts, anchor bolt bundles or anchor cables applied to a collapse-type block should be arranged in the vertical or nearly vertical direction, and the force shall be projected to the vertical direction when calculating the supporting force. When slipping-type block is reinforced with the prestressed anchor bolts or anchor cables, the direction of the anchor bolts or anchor cables should be determined considering the sliding direction of the block and construction conditions.

7.3.4 The anchor bolts, anchor bolt bundles and anchor cables shall be driven into stable surrounding rock, and the length of the anchored section shall meet the requirements of the pull-out resistance capacity. The calculation of the pull-out resistance capacity of the anchored section shall comply with the current national standard GB 50086, *Technical Code for Engineering of Ground Anchorages and Shotcrete Support*.

7.3.5 For the tunnel with developed fractures in surrounding rock or the alignment failing to satisfy Item 1 of Article 4.1.2 of this code, systematic anchor bolts or anchor bolt bundles should be adopted, and the arrangement should meet the following requirements:

1 Anchor bolts or anchor bolt bundles should be arranged perpendicular to the major discontinuities. If the orientation of the major discontinuities is indiscernible, the anchor bolts may be arranged normal to the peripheral outline of the tunnel.

2 Anchor bolts or anchor bolt bundles should be staggered on the surrounding rock surface.

3 The spacing between anchor bolts or bundles should not be more than 1/2 of its length; for the tunnel section with poor surrounding rock, the spacing should not exceed 1.0 m.

7.3.6 The tunnel with high in situ stress should be reinforced by pre-stressed anchor bolts or grouted anchor bolts with face plate. The tunnel prone to failure of surrounding rock or requiring greater support should be reinforced by pre-stressed anchor bolts or anchor cables.

7.4 Anchor-Shotcrete Support with Steel Mesh

7.4.1 The surrounding rock with cataclastic rocks and developed fractures should adopt anchor-shotcrete support with steel mesh.

7.4.2 The layout of the reinforcing mesh shall meet the following requirements:

1 The bar diameter of steel mesh should be 6 mm to 12 mm, and the spacing should be 0.15 m to 0.20 m.

2 The steel mesh and anchor bolts should be connected by welding. The intersection points of steel mesh should be welded and tied alternatively.

7.4.3 The thickness of shotcrete with steel mesh should not be less than 0.10 m, and the minimum thickness of concrete cover should be 0.05 m.

7.4.4 For the tunnel section with poor geological conditions where the surrounding rock is extremely unstable, the combined support of anchor-shotcrete with steel mesh and steel rib may be adopted.

7.5 Combined Support

7.5.1 Combined support shall be composed of primary support and secondary support. Anchor bolts-shotcrete, steel mesh, steel rib or their combinations may be adopted for primary support. Concrete lining or reinforced concrete lining may be adopted for secondary support.

7.5.2 The layout and support strength of primary support shall be coordinated with secondary support. According to monitoring measurement, if the primary support meets the stability requirements of the surrounding rock, the surrounding rock pressure acting on the secondary support may be neglected or lessened in the structural calculation.

8 Basic Principles for Structural Design

8.0.1 Probability-based design at ultimate limit state by partial factors shall be adopted for the structural design of hydraulic tunnels.

8.0.2 The design of hydraulic tunnel structure shall be conducted based on persistent, transient and accidental design situations.

8.0.3 The limit state design shall meet the following requirements:

1 In all design situations, ultimate limit state design shall be carried out.

2 In the persistent design situation, serviceability limit state design shall be carried out.

3 In the transient design situation, serviceability limit state design may be carried out as required.

4 In the accidental design situation, serviceability limit state design need not be carried out.

8.0.4 In the ultimate limit state design, the combinations of action effects shall be adopted according to the design situations and the following requirements:

1 Fundamental combination, which consists of permanent and variable action effects, shall be adopted for persistent and transient design situations.

2 Accidental combination, which consists of permanent, variable and accidental action effects, shall be adopted for accidental design situations. Only one accidental action shall be considered for each accidental combination.

8.0.5 In serviceability limit state, characteristic combination, or characteristic combination coupled with long-term action effect shall be adopted.

8.0.6 The structure safety class and importance factor of hydraulic tunnel shall be determined as per Table 8.0.6. For the hydraulic tunnel of great importance, the structure safety class shall be determined after demonstration.

Table 8.0.6 Structure safety class and importance factor of hydraulic tunnel

Grade of hydraulic structure	Structure safety class of hydraulic tunnel	Importance factor of structure γ_0
1	Ⅰ	1.1
2, 3	Ⅱ	1.0
4, 5	Ⅲ	0.9

9 Concrete and Reinforced Concrete Lining

9.1 General Requirements

9.1.1 The hydraulic tunnel should be lined with concrete or reinforced concrete in any of the following cases:

1 The surface of surrounding rock needs to be smoothed to reduce the head loss.

2 The scour by water flow and damages to surrounding rock by water, air and the change of temperature and humidity need to be prevented.

3 The seepage resistance of the tunnel needs to be enhanced.

4 The primary support fails to meet the long-term stability requirements of the tunnel.

5 The deep covered tunnel is located in the region with high in situ stress.

9.1.2 For the tunnel with Classes Ⅰ, Ⅱ or part of Class Ⅲ surrounding rock which meets the requirements of Items 1 and 2 of Article 9.1.1 of this code, the concrete lining without considering bearing load should be adopted and its water-facing side should be reinforced according to detailing requirements.

9.1.3 For the high pressure tunnel mainly with Classes Ⅱ and Ⅲ surrounding rock which meets the requirements of Items 1, 2 and 3 of Article 4.1.4 of this code, reinforced concrete lining may be adopted after techno-economic analysis.

9.1.4 When the reinforced concrete lining of the tunnel fails to meet the requirements of safety and serviceability, the following measures should be taken:

1 Consolidation grouting of surrounding rock.

2 Steel lining backfilled with concrete.

3 Prestressed concrete lining.

9.1.5 The thickness of concrete lining or reinforced concrete lining should be determined according to the calculation and analysis with reference to detailing requirements and the construction method, and may be preliminarily taken as 1/16 to 1/12 of the inner diameter. The minimum thickness of the concrete lining with single-layer reinforcing steel bars should be 0.3 m and that of the concrete lining with double-layer reinforcing steel bars should be 0.4 m.

9.1.6 The thickness of concrete cover on the flow surface of the high flow velocity tunnel should not be less than 0.10 m, and the reinforcing steel bars on the surface should be parallel to the flow direction. The thickness of concrete cover elsewhere should not be less than 0.05 m.

9.1.7 The hydraulic tunnel shall meet the durability requirements of structure. The minimum strength class, frost resistance grade and impermeability grade of lining concrete shall be selected in accordance with the current sector standard DL/T 5057, *Design Specifications for Hydraulic Concrete Structures*, considering environmental conditions, design service life, structural type, climatic conditions, operation conditions of the tunnel, etc.

9.1.8 Allowable crack design may be adopted when tunnel lining is designed as per ultimate limit state. When the tunnel lining is designed as per serviceability limit state, characteristic combination coupled with the effects of long-term actions shall be considered to check the crack width, and the maximum allowable crack width shall comply with the current sector standard DL/T 5057, *Design Specifications for Hydraulic Concrete Structures*.

9.1.9 For the tunnel without considering seepage resistance, the serviceability limit state calculation need not be performed.

9.1.10 In the structural calculation of reinforced concrete lining, the structural factor γ_d shall be taken as follows:

1 For circular pressure tunnels, γ_d shall be taken as 1.35 when using the calculation method in Appendix D of this code.

2 For circular free-flow tunnels, inverted U-shaped tunnels, horse-shoe-shaped tunnels, and tunnels with other section shapes, γ_d shall be selected in accordance with the current sector standard DL/T 5057, *Design Specifications for Hydraulic Concrete Structures*.

9.2 Actions and Combination of Action Effects

9.2.1 Actions and their partial factors shall be adopted as per Table 9.2.1-1. The combination of action effects at ultimate limit state shall be determined as per Table 9.2.1-2, and the combination of action effects at serviceability limit state shall be determined as per Table 9.2.1-3.

9.2.2 The calculation of surrounding rock pressure and in situ stress shall comply with the current national standard GB/T 51394, *Standard for Load on Hydraulic Structures*. The seismic action calculation and seismic safety check of tunnel structure shall comply with the current sector standard NB 35047, *Code for Seismic Design of Hydraulic Structures of Hydropower Project*.

Table 9.2.1-1 Actions and corresponding partial factors

Type of action	Action	Partial factor
Permanent action	Surrounding rock pressure, in situ stress	1.0
	Dead weight of lining	1.1 (0.9)
Variable action	Hydrostatic pressure under normal operation condition	1.0
	Water hammer pressure, surge pressure	1.1
	Fluctuating pressure	1.0
	Groundwater pressure	1.0
	Backfill grouting pressure	1.3
Accidental action	Hydrostatic pressure at check flood level	1.0
	Seismic action	1.0

NOTE When the effect of dead weight action is beneficial to the structure, the partial factor of action shall be taken as the value in brackets in the table.

Table 9.2.1-2 Combination of action effects at ultimate limit state

Design situation	Combination of action effects	Main conditions	Types of action							
			Surrounding rock pressure, in situ stress in rock mass	Dead weight of lining	Hydro-static pressure	Water hammer pressure, surge pressure	Fluctuating pressure	Ground-water pressure	Backfill grouting pressure	Seismic action
Persistent situation	Fundamental combination	Normal operation of pressure conduit for conventional hydropower station	√	√	√	√	√	√	–	–
		Normal operation for flow release tunnel	√	√	√	–	√	√	–	–
		Normal operation of pressure conduit for pumped storage power station	√	√	√	√	√	√	–	–
Transient situation	Fundamental combination	Construction period of tunnel	√	√	–	–	–	√	√	–
		Maintenance period of tunnel	√	√	–	–	–	√	–	–

Table 9.2.1-2 *(continued)*

Design situation	Combination of action effects	Main conditions	Types of action							
			Surrounding rock pressure, in situ stress in rock mass	Dead weight of lining	Hydro-static pressure	Water hammer pressure, surge pressure	Fluctuating pressure	Ground-water pressure	Backfill grouting pressure	Seismic action
Accidental situation	Accidental combination	Operation of pressure conduit for conventional hydropower station at check flood level	√	√	√	√	√	√	–	–
		Operation of flow release tunnel at check flood level	√	√	√	–	√	√	–	–
		Operation of pressure conduit for pumped storage power station at check flood level	√	√	√	√	√	√	–	–
		Earthquake	√	√	√	√	–	√	–	√

NOTE The fluctuating pressure of pressure conduit of a conventional hydropower station may be determined according to the specific operation conditions of the project.

Table 9.2.1-3 Combination of action effects at serviceability limit state

<table>
<tr><th rowspan="2">Design situation</th><th rowspan="2">Combination of action effects</th><th rowspan="2">Main conditions</th><th colspan="7">Types of action</th></tr>
<tr><th>Surrounding rock pressure, in situ stress in rock mass</th><th>Dead weight of lining</th><th>Hydrostatic pressure</th><th>Water hammer pressure, surge pressure</th><th>Fluctuating pressure</th><th>Groundwater pressure</th><th>Backfill grouting pressure</th></tr>
<tr><td rowspan="3">Persistent situation</td><td rowspan="3">Characteristic combination</td><td>Normal operation of pressure conduit for conventional hydropower station</td><td>√</td><td>√</td><td>√</td><td>√</td><td>√</td><td>√</td><td>–</td></tr>
<tr><td>Normal operation of flow release tunnel</td><td>√</td><td>√</td><td>√</td><td>–</td><td>√</td><td>√</td><td>–</td></tr>
<tr><td>Normal operation of pressure conduit for pumped storage power station</td><td>√</td><td>√</td><td>√</td><td>√</td><td>√</td><td>√</td><td>–</td></tr>
<tr><td rowspan="2">Transient situation</td><td rowspan="2">Characteristic combination</td><td>Construction period of tunnel</td><td>√</td><td>√</td><td>–</td><td>–</td><td>–</td><td>√</td><td>√</td></tr>
<tr><td>Maintenance period of tunnel</td><td>√</td><td>√</td><td>–</td><td>–</td><td>–</td><td>√</td><td>–</td></tr>
</table>

NOTE The fluctuating pressure of pressure conduit for conventional hydropower station may be determined according to the project-specific operation conditions.

9.2.3 The actions of surrounding rock with special properties such as rheology, swelling and softening due to water absorbing shall be determined after demonstration.

9.2.4 The internal water pressure of the tunnel should be determined according to the characteristic water levels of the inlet and outlet considering the project-specific conditions.

9.2.5 For the pressure tunnel with a surge chamber, the characteristic value of internal water pressure at the calculated section should be determined as follows:

1 Under fundamental combination or characteristic combination considering long-term effect, the characteristic value of internal water pressure in the tunnel with a conventional surge chamber should be determined by linear interpolation from the normal water level at the inlet or outlet to the corresponding maximum surge water level. For the tunnel with an air cushion surge chamber, the characteristic value of internal water pressure should be determined by linear interpolation from the normal water level at the inlet to the sum of corresponding maximum surge water level and the water head equivalent to air pressure.

2 Under accidental combination, the characteristic value of internal water pressure at the calculation section of the tunnel with a conventional surge chamber should be determined by linear interpolation from the highest water level at the inlet or outlet to the corresponding maximum surge water level. For the tunnel with an air cushion surge chamber, the characteristic value of internal water pressure should be determined by linear interpolation from the highest water level at the inlet to the sum of corresponding maximum surge water level and the water head equivalent to air pressure.

9.2.6 The external water pressure on concrete lining may be determined according to the measured data or as per Appendix E of this code. For the deep-covered tunnel and tunnel with complex hydrogeological conditions, the external water pressure should be determined by special seepage field analysis.

9.2.7 The influence of temperature change, stresses caused by drying shrinkage and expansion of concrete and consolidation grouting pressure of surrounding rock on lining should be handled by detailing measures and construction measures. The temperature stresses in high geothermal area shall be studied.

9.2.8 The construction load may be determined according to the mechanical force during construction or maintenance.

9.2.9 The influence of inclination angle of shaft axis shall be considered in the calculation of the dead weight of lining, pressure of surrounding rock and full water pressure in the tunnel, which may be calculated as per the component force perpendicular to shaft axis.

9.3 Lining Calculation

9.3.1 The structural calculation of load-bearing lining shall be carried out according to the tunnel function, size, lining type, surrounding rock conditions, construction method, etc., and shall meet the following requirements:

1 The finite element method should be adopted for high-pressure tunnels or hydraulic tunnels whose inner diameter (net span) are not less than 10 m.

2 For the circular pressure tunnel, if the lining mainly bears the internal water pressure, the surrounding rock is relatively homogeneous, and the cover thickness meets the requirements of Items 1 and 2 of Article 4.1.4 of this code, the lining may be calculated by the formulae given in Appendix D of this code, where the elastic resistance of surrounding rock shall be considered. When the thickness of the surrounding rock of the tunnel is less than 3 times the excavation diameter, the elastic resistance value shall be demonstrated.

3 The stresses of circular tunnel lining bearing uniform external water pressure may be calculated as per Appendix F of this code, and the circumferential stresses shall be less than the allowable design value of concrete stresses.

4 Boundary-value numerical method should be adopted for free-flow tunnels and non-circular pressure tunnels.

9.3.2 The lining which bears asymmetric load may be specifically calculated based on topographical and geological conditions.

9.3.3 For multi-parallel tunnels, the stress change of surrounding rock caused by excavation of adjacent tunnels shall be considered in lining calculation, and the finite element method should be adopted.

9.3.4 The finite element analysis of seepage and bearing structure should be carried out for the high-pressure concrete-lined tunnels or concrete-lined tunnels in important projects. The lining thickness and reinforcement may be determined considering the calculation results, engineering analogy and detailing requirements comprehensively.

9.3.5 The seismic design of lining structure shall comply with the current sector standard NB 35047, *Code for Seismic Design of Hydraulic Structures of Hydropower Project*.

9.4 Lining Joint

9.4.1 Deformation joints shall be provided at the locations where the geological conditions change, the tunnel/shaft and inlet/outlet structure are connected, and relatively large deformation may occur. Bellows expansion joints may be set where the low-velocity tunnel passes through the active fault or the boundary of bedrock and overburden, and corresponding seepage control measures may be taken. For the deformation joint with small temperature change in the tunnel, the joint width and joint filler need not be set, and emulsified asphalt should be painted on the joint surface.

9.4.2 For tunnel sections with homogenous surrounding rock conditions, only construction joints may be provided. The construction joints spacing may be determined according to the construction method, concrete placement capacity and air temperature variation, and should be taken as 6 m to 12 m. The circumferential joints at the bottom (invert), sidewall and crown should not be staggered.

9.4.3 Waterstops need not be set for the deformation joints and circumferential construction joints for the low-velocity tunnel without seepage control requirements, and the longitudinal reinforcement need not pass through the construction joint surface. Reliable waterstops shall be set for the deformation joints and circumferential construction joints for the tunnel with high flow velocity or seepage control requirements, the longitudinal reinforcement should pass through the construction joint surface, and the seepage control and joint treatment measures shall be taken according to specific conditions.

9.4.4 No joints shall be set at the connection between the reinforced concrete lining and steel lining. The steel lining shall extend into the reinforced concrete lining for no less than 1.0 m, and water resistance measures shall be taken at the connection.

9.4.5 Longitudinal construction joints shall be set at the locations with small tensile stresses of the lining structure. Where the sidewall and crown are firstly lined, reliable treatment shall be conducted for the inverted joints.

10 Prestressed Concrete Lining

10.1 General Requirements

10.1.1 For the tunnels with higher requirements for seepage control or with overlying rock mass inadequate to prevent hydraulic fracturing, prestressed concrete lining may be adopted based on techno-economic comparison.

10.1.2 The prestressing in concrete lining may be of the grouting type or circular anchoring type. The lining type should be selected based on geological conditions, construction conditions and operation requirements. The grouting-type prestressed lining may be used for the tunnels with overlying rock mass adequate to prevent hydraulic fracturing, otherwise the circular anchoring type should be adopted.

10.1.3 Circular section shall be adopted for the prestressed concrete lining. Smooth blasting should be used for excavation, and backfill concrete should be used for restoration first in the case of large overbreaks.

10.1.4 The lining thickness shall be determined by calculation based on load combinations in different conditions, which should be 1/18 to 1/12 of the tunnel diameter. The minimum lining thicknesses should be 0.6 m for the circular anchoring type and 0.3 m for the grouting type. The structural calculation of the lining shall meet the following requirements:

1 Under the combined action of internal water pressure, prestress and other loads, the tensile stress in the lining shall be less than the allowable tensile stress of the concrete.

2 Under the combined action of prestress and other loads without internal water pressure, the compressive stress in the lining shall be less than the allowable compressive stress of the concrete.

10.1.5 The design indexes of the material properties of concrete and prestressed reinforcement shall be in accordance with the current sector standard DL/T 5057, *Design Specification for Hydraulic Concrete Structures*.

10.1.6 When prestress is applied, the required cubic concrete compressive strength shall be determined by calculation, but should not be lower than 75 % of the design strength value of the concrete cube under the same curing conditions.

10.1.7 Prestressed concrete lining shall be subject to ultimate limit state calculation and serviceability limit state verification.

10.2 Grouting-Type Prestressed Concrete Lining

10.2.1 The grouting pressure shall be determined based on the principle that there is no tensile stress in the lining under the maximum internal water pressure. The grouting pressure should not be less than 2 times the maximum internal water pressure, and expanding cement should be used for the grout.

10.2.2 Grouting holes shall be evenly distributed along the lining perimeter, with a spacing of 2 m to 4 m. For tunnels with a diameter less than 5 m, 8 to 10 holes should be provided for every row. And for tunnels with a diameter of 5 m to 10 m, 8 to 12 holes may be provided. The length of grouting sections should be 2 to 3 times the tunnel diameter.

10.2.3 The construction process and grouting parameters shall be determined by tests. The grouting sequence should be as follows:

1 Conduct consolidation grouting of surrounding rock.

2 Inject high-pressure water between the surrounding rock and the concrete lining until they are completely separated from each other.

3 Conduct high-pressure grouting between the surrounding rock and the concrete lining.

10.3 Circular Anchored Prestressed Concrete Lining

10.3.1 The circular anchored lining may include bounded and unbounded post-tensioned prestressed linings, and the priority should be given to the latter in design.

10.3.2 The design parameters of prestressed concrete lining shall be determined by tests. The governing tensioning stress σ_{con} of prestressed reinforcement shall be adopted in accordance with the current sector standard DL/T 5057, *Design Specification for Hydraulic Concrete Structures*, according to the types of prestressing reinforcement, and tensioning methods.

10.3.3 Prestressed reinforcement shall be placed at the outer edge of the lining, and the spacing shall be determined by calculation, but should not be greater than 0.5 m.

10.3.4 The positions of anchorages should be staggered.

10.3.5 The parameters and construction process of circular-anchored lining shall be determined by tests. The grouting shall meet the following requirements:

1 The full-face contact grouting shall be conducted between the lining and the surrounding rock.

2 For the bounded post-tensioned prestressed lining, the hole-fill grouting and tensioning-slot backfilling shall be conducted after the completion of anchor cable tensioning.

11 High-Pressure Bifurcation Tunnel with Reinforced Concrete Lining

11.0.1 The high-pressure reinforced concrete-lined bifurcation tunnel should be adopted through techno-economic verification according to the geological conditions, in situ stress and permeability of surrounding rock, hydraulic stability, construction conditions, etc.

11.0.2 The structural safety class of the high-pressure reinforced concrete-lined bifurcation tunnel shall be identical to that of the main tunnel.

11.0.3 The location and outline of the high-pressure reinforced concrete-lined bifurcation tunnel shall be determined through comprehensive analysis and demonstration according to the project layout, hydraulic conditions, surrounding rock conditions, structure load conditions, etc.

11.0.4 If a high-pressure reinforced concrete-lined bifurcation tunnel is adopted, the surrounding rock shall meet the following requirements:

1 Class I and Class Ⅱ surrounding rock with slight permeability should be predominate.

2 The minimum initial in situ stress of the surrounding rock shall be greater than the design hydrostatic pressure at this position of the tunnel, and the hydraulic fracturing shall not occur under the action of internal water pressure.

3 The top and side cover thickness of rock mass of the bifurcation tunnel shall meet the requirements of Article 4.1.4 of this code.

4 The clear distance between the bifurcation tunnel and adjacent caverns shall be determined according to the seepage stability requirements.

11.0.5 In the design of the high-pressure reinforced concrete-lined bifurcation tunnel located in Class Ⅰ and Class Ⅱ surrounding rock with slight permeability, the action of internal water pressure on the reinforced concrete lining may be ignored. Therefore, the reinforcement configuration should be determined by engineering analogy and according to detailing requirements.

11.0.6 The structural design for high-pressure reinforced concrete-lined bifurcation tunnels with complex conditions or bifurcation tunnels of important projects shall be carried out in accordance with Article 9.3.4 of this code.

11.0.7 The excavation and support procedures shall be strictly controlled in bifurcation tunnel construction.

12 Design for Machine-Bored Tunnel

12.0.1 The tunnel boring machine (TBM) may be used for the construction of long tunnel projects through techno-economic comparison.

12.0.2 The following geological factors shall be considered for the layout of machine-bored tunnels:

1 Topographical and geomorphologic conditions along the tunnel alignment, and the stability conditions of the inlet and outlet slopes.

2 Distribution of strata and lithology, attitude of rock strata along the tunnel alignment, locations, scales and properties of faults, fractured zones and densely jointed and fissured zones, and distribution of in situ stress field along the tunnel alignment.

3 Classification of surrounding rock, physical and mechanical properties and parameters of various rock masses.

4 Groundwater table, water temperature and chemical composition, aquifer, water catchment structures and predicted water burst volume.

5 Hazardous gases and radioactive elements in the tunnel.

12.0.3 A machine-bored tunnel should be as straight as possible, and the bend radius shall meet the requirements of the minimum turning radius of the selected TBM. The areas where the geological conditions are unsuitable for the TBM construction should be avoided in the layout of tunnel alignment.

12.0.4 The longitudinal slope of machine-bored tunnel shall be determined according to the requirements of tunnel functions, construction drainage, construction equipment performances and transportation mode.

12.0.5 A machine-bored tunnel shall have a circular section, and the section size shall be determined according to the tunnel function requirements, thickness of support and lining, deformation of surrounding rock, considering the factors such as cut deviation, cutter head abrasion, etc., and shall meet the requirements of the minimum excavation size of the TBM.

12.0.6 The type of TBM shall be selected according to geological conditions, lining style of the tunnel, etc.

12.0.7 The support design for a machine-bored tunnel shall meet the requirements of surrounding rock stability and construction performances of the TBM.

12.0.8 The design of lining structure and grouting for tunneling by an open

TBM may follow the requirements of tunneling by drilling-and-blasting. For the tunnel lined with segment-assembled concrete, adjacent segments shall be tightly connected. The interspace between the segments and surrounding rock shall be backfilled with pea gravel and grouted tightly. The design parameters of segments should be determined by the finite element calculation method.

12.0.9 The construction of a machine-bored tunnel shall follow the principle of "exploration prior to excavation", and advance geological prediction may be carried out by a combination of multiple methods.

12.0.10 The tunneling speed of the TBM shall be slowed down and the treatment measures shall be taken in advance when tunneling in poor geological conditions. When special geological condition is encountered, the tunneling should be stopped and may not be continued until engineering measures are taken to ensure the safety, or the tunnel section is treated by other measures.

12.0.11 The size of auxiliary cavern for the machine-bored tunnel shall meet the requirements of installation, disassembly and safe passing of the TBM during construction, and engineering measures shall be taken to meet the operation requirements of the tunnel.

13 Design for Special Rock Mass and Poor Geological Tunnel

13.0.1 For special rock mass and tunnel sections with poor geological conditions such as fault fracture zone, weak broken surrounding rock, water-enriched stratum, rock burst or hazardous gas, karst cave and swelling stratum, etc., special design shall be conducted according to the specific conditions, and corresponding engineering measures shall be taken. During construction, the observation of the surrounding rock and groundwater table variation, monitoring of support and lining shall be strengthened, and the support measures shall be modified in time.

13.0.2 The support design for tunnel sections with poor geological conditions shall meet the following requirements:

1 The reinforcement design of the surrounding rock before support or excavation shall be determined by engineering analogy, calculations and analysis based on the geological prediction/forecast or advance exploration results.

2 For the support scheme, the support parameters shall be confirmed, adjusted, or modified timely according to the revealed geological conditions, site monitoring and test data during construction, to prevent surrounding rock instability or its worsening.

3 The effect of initial support shall be analyzed timely according to the stability of surrounding rock, and the necessity of reinforced support or multi-support shall be studied.

4 For possible unexpected circumstances in the tunnel sections with poor geological conditions, an emergency plan shall be prepared.

13.0.3 The lining design for the tunnel sections with poor geological conditions shall meet the following requirements:

1 The loads imposed on the lining structure should be determined by engineering analogy and calculation according to the geological conditions, the effects of various treatment measures taken before lining construction and the stability of surrounding rock deformation.

2 The surrounding rock's physical and mechanical indexes and the ability to bear internal water pressure adopted in design shall be determined by the relevant tests and engineering analogy.

3 The tunnel section shape and lining type favorable to the surrounding rock stability and structure stress shall be selected through techno-economic comparisons according to the geological and construction conditions.

4 For the lining structure calculation of the tunnel sections with poor geological conditions, structural mechanics methods may be used when the internal water pressure acting on surrounding rock is ignored; otherwise, the finite element method may be used, where engineering analogy may also be used.

13.0.4 For the tunnel sections where collapse failure of surrounding rock may occur according to geological prediction/forecast, the tunneling shall be conducted using the New Austrian Tunneling Method (NATM), and the following requirements shall be met:

1 Specific construction planning should be proposed.

2 The technical requirements for construction should be proposed, including blasting parameters, footage, procedure, deformation monitoring, field tests, support process, etc.

3 The diversion and drainage of groundwater should be designed.

4 The stability of the surrounding rock shall be timely assessed on the basis of geological information feedback during construction to determine the subsequent construction measures.

13.0.5 For the tunnel sections with poor geological conditions in water-enriched strata, measures to prevent water inrush, water drainage and diversion measures, or support measures shall be studied in design. Construction monitoring, lining structure design and safety monitoring shall be carried out, and the following requirements shall be met:

1 For the tunnel sections with major water burst, engineering measures, such as cutting off water sources, diverting and draining water and reducing rock permeability by grouting, shall be taken according to the geological conditions, sources and inflow of water.

2 For the tunnel sections with high-pressure water burst, the measures such as pressure reduction by diverting partial flow and zoned treatment shall be taken according to the pressure, sources and inflow of water, etc.

3 For the tunnel sections in water-enriched strata, engineering measures to prevent or control water burst from causing instability of surrounding rock shall be taken.

13.0.6 The design of tunnel sections where rock burst occurs in high in situ stress areas shall meet the following requirements:

1 For tunnel sections where rock burst occurs in high in situ stress areas, the orientation and cross-sectional shape of the tunnel, excavation procedure, support methods, prerelease of surrounding rock pressure, etc. shall be studied according to the magnitude and direction of in situ stress, lithology and structure of the surrounding rock, and frequency, intensity and extent of rock burst, to prevent further development of rock burst.

2 The combined support of shotcrete, systematic anchor bolts and steel mesh may be used for the initial support of rock burst section. Where the rock burst intensity is high, measures such as advance stress release, preset anchor bolt, steel ribs and zoned support may be taken.

3 The effects of primary support shall be monitored closely. Lining construction shall be carried out after the surrounding rock deformation is basically converged.

13.0.7 For tunnel sections through areas with hazardous gases, measures such as isolation, closure, ventilation and exhausting shall be taken according to the sources, distribution and connectivity of hazardous gases to control the impacts of hazardous gases. For tunnels with a large length or excessive concentration of hazardous gases, ventilation shall be strengthened during construction and maintenance periods. Anchor-shotcrete support should not be used as permanent support in hazardous gas areas.

13.0.8 For tunnels through karst or cave areas, according to the location, distribution, size and filling status of karst caves, stability of the surrounding rock and groundwater conditions, the following treatment measures shall be taken:

1 For the water seeping or dripping from rock wall, flowing water in karst caves and groundwater in fills, comprehensive treatment measures such as drainage, cutting off, blocking and prevention should be taken according to the water quantity, type and sources.

2 Treatment measures such as backfill concrete, backfill grouting and consolidation grouting may be taken for relatively small karst caves or small karst caves not connecting with tunnels.

3 For large-sized, fills-abundant or water-enriched karst caves, treatment measures such as partition, supporting structure for crossing, specialized foundation and local alignment modification may be taken according to the location and distribution of caves.

4 For the cavern wall with an insufficient strength or unstable conditions that might affect the safety of the tunnel, measures such as propping, anchorage and grouting shall be taken.

13.0.9 For tunnel sections crossing the strata with high swelling and rheologic rock, the swelling ratio and swelling pressure of the swelling rock, and the timeliness and stress/strain relationship of rheologic rock should be studied based on geological exploration and test results. The working face enclosing modes and time, support measures, lining structure type and its construction period shall be determined through engineering analogy, calculation and analysis.

13.0.10 For tunnel sections with loose deposit layer and aqueous sand layer, and other weak layers that are prone to argillitization, disintegration, swelling or softening in water, or with large faults, unloading zones, fracture zones, zones with dense joints and fissures that are prone to alteration and seepage deformation failure under seepage action, the structural safety shall be upgraded by one level but shall not exceed Class Ⅰ. The design of seepage control and waterstop for lining structure shall meet the following requirements:

1 The cross-sectional shape of the tunnel section with weak and swelling surrounding rock should be circular or nearly circular.

2 For tunnel sections in loose deposit layer or aqueous sand layer, the pre-reinforcement measures such as surface grouted anchor bolt, and grouting from the surface or the tunnel periphery to the surrounding rock should be taken before construction. Advance supporting measures such as preset anchor bolt, advance small grouted pipe or pipe shed may be adopted during construction.

3 The surface water and groundwater that might affect the safety of tunnel structure shall be treated according to the specific situation.

13.0.11 For tunnel sections with poor geological conditions, the backfill grouting, consolidation grouting, waterproofing and drainage, waterstop for construction joints and structure joints, as well as safety monitoring combined with construction monitoring shall be well designed according to the geological conditions and lining type.

14 Plugging Body Design

14.1 General Requirements

14.1.1 The plugging locations of hydraulic tunnels shall be analyzed and determined according to engineering geological and hydrogeological conditions of surrounding rock, layout of adjacent structures, tunnel supporting and lining conditions and operation requirements, and should be arranged in the tunnel section with relatively good geological conditions.

14.1.2 For the hydraulic tunnel plug directly contacting reservoir water, its structure grade and stability and seepage control requirements shall be the same as those of water retaining structures. The structure grade of adit plugs shall be the same as that of the main tunnels.

14.1.3 The plug shape shall be selected according to the cross-sectional shape of hydraulic tunnels, construction conditions, engineering geological conditions, etc. A plug with wedge-shaped longitudinal section is preferred for high-pressure or large-section tunnels.

14.1.4 The layout of the tunnel plug shall meet the following requirements:

1 When the tunnel alignment passes through the grout curtain line of water retaining structure, the plug shall be located on the curtain line and connected with the curtain. When enlarged excavation needed for the plug, reinforced curtain grouting shall be performed to the perimeter surrounding rock.

2 For the hydraulic tunnels, such as flood discharge tunnel and emptying tunnel, which are transformed from diversion tunnels, the layout of plugs should take account of the transformation of the hydraulic tunnel, and shall meet the structural loading and seepage control requirements.

14.2 Design and Calculation

14.2.1 Plugs shall be of concrete structure, the strength should not be inferior to C20 for upstream face, and not be inferior to C15 for other parts.

14.2.2 Plugs shall be designed on the basis of the ultimate limit states. The actions and their partial factors, and action effects combination shall be in accordance with Article 9.2.1 of this code.

14.2.3 The structural factor of plug γ_d shall be taken in accordance with the current sector standard DL/T 5057, *Design Specification for Hydraulic Concrete Structures*.

14.2.4 The stability against sliding of columnar plugs may be calculated by the following formulae:

$$S(\cdot) = \Sigma P_R \tag{14.2.4-1}$$

$$R(\cdot) = f_R \Sigma W_R + C_R (A_{R1} + \lambda A_{R2}) \tag{14.2.4-2}$$

where

$S(\cdot)$ is the function of action effects;

$R(\cdot)$ is the function of resistance;

ΣP_R is the sum of all tangential actions of the plug on the sliding plane (kN);

ΣW_R is the sum of all normal actions of the plug on the sliding plane (kN), taken as positive for the gravitational direction;

f_R is the friction coefficient between concrete and surrounding rock or at concrete interface;

C_R is the cohesion between concrete and surrounding rock or at concrete interface (kPa);

A_{R1} is the contact area of plug bottom with surrounding rock (m^2);

A_{R2} is the contact area of plug sides with surrounding rock (m^2);

λ is the area coefficient of the effective contact surface of plug sides with surrounding rock, taken as 0.3 to 0.8 according to the project-specific conditions.

14.2.5 In addition to meeting the requirements of stability against sliding, the plug length of a high-pressure tunnel shall meet the seepage stability requirements. The length may be estimated by the following formula:

$$\frac{H}{L} \leq [k] \tag{14.2.5}$$

where

H is the design water head (m);

L is the length of plug (m);

$[k]$ is the allowable hydraulic gradient for bypass seepage of surrounding rock.

14.2.6 The finite element analysis should be conducted for the plug with high internal water pressure. Seepage analysis should be conducted with the finite element method for the plug with complex surrounding rock conditions and directly contacting reservoir water.

14.3 Structural Requirements

14.3.1 The excavation outline of plug should be completed in one round together with the excavation of the main tunnel. Second excavation should be avoided.

14.3.2 Anchor bars should be provided between the plug and the surrounding rock. The spacing of anchor bars should not be greater than 2 m. The anchor bars may be inserted into the surrounding rock to a depth of 2 m to 4 m, and their length in the plug should not be less than 0.5 m.

14.3.3 The micro-expansive concrete may be adopted as plug materials, and the type and dosage of the expansive admixture should be determined by tests. The backfill grouting must be performed at the top of the plug, and the contact grouting at the plug circumference should be determined by the plug location, water head and importance. The consolidation grouting of surrounding rock at the plugging section should be determined according to the engineering geological conditions and seepage control requirements. The spacing of consolidation grouting holes should be 2 m to 3 m, and the grouting holes should penetrate into the surrounding rock to a depth not less than 0.5 times the diameter (or span) of the excavated tunnel. A grouting gallery should be provided in the plug.

14.3.4 The consolidation grouting for the plugging section of the diversion tunnel should be completed before the river closure. The joint grouting, contact grouting and reinforcement curtain grouting should be performed in grouting galleries and shall be completed before reservoir impoundment.

14.3.5 The waterstop and jointing shall be well designed for the connection between the plug of adit and the lining structure of main tunnel with due consideration of structural loading and construction method.

14.3.6 Transverse construction joints need not be set for the plug with a length less than 20 m. The concreting lifts of the plug shall be determined according to the concrete placing capacity, construction methods, temperature control requirement, etc. The construction joints shall be roughened and dowels shall be installed. Joint grouting shall be carried out between the plug and the previously placed tunnel concrete lining. The joint grouting shall not be carried out until the concrete temperature gets stable.

14.3.7 For a pressure diversion tunnel, the longitudinal excavated triangle part reserved in the plugging section of the main tunnel should be backfilled temporarily prior to the river closure.

15 Grouting and Seepage Control and Drainage

15.1 Grouting

15.1.1 The backfill grouting shall be conducted between the top of tunnel concrete, reinforced concrete lining, or plug and the surrounding rock.

15.1.2 The method, range, hole spacing, row spacing, pressure and grout concentration of backfill grouting shall be determined through analysis according to the structural types, operation conditions and construction methods of the lining or plugging. The backfill grouting range should be at the top of the tunnel or within 90° to 120° central angle of the crown; and for other portions, the range depends on the placing conditions of lining concrete. The grouting hole spacing and row spacing should be 2 m to 4 m, and the grouting pressure should be 0.2 MPa to 0.3 MPa. The diameter of grouting holes should not be less than 38 mm, and the grouting depth shall not be less than 0.3 m into holes.

15.1.3 Cement blocks formed by backfill grouting shall meet the requirements of resistance transfer.

15.1.4 The consolidation grouting of the surrounding rock shall be determined by techno-economic comparison according to the engineering geological conditions, hydrogeological conditions and tunnel operation requirements. Consolidation grouting parameters shall meet the following requirements:

1 Consolidation grouting parameters may be determined by engineering analogy or site tests. The row spacing for grouting holes should be 2 m to 4 m, and each row should have at least 6 holes. Adjacent rows should be staggered. The grouting depth into the surrounding rock shall be determined according to the surrounding rock conditions, which should be not less than 0.5 times the diameter or the width of the tunnel. The grouting pressure should be 1 to 2 times the hydrostatic pressure.

2 High-pressure consolidation grouting parameters of high-pressure reinforced concrete lined tunnels and bifurcation tunnels may be determined by zones according to the change of internal water pressure. The grouting pressure shall be 1.2 to 1.5 times the hydrostatic pressure, and shall be less than the minimum principal stress of the surrounding rock. The grouting depth into the surrounding rock should be 0.5 to 0.75 times the tunnel diameter (or span). The parameters shall be determined reasonably by consolidation grouting tests and water pressure tests.

3 The permeability of surrounding rock shall not be greater than 1 Lu for high-pressure reinforced concrete bifurcation tunnel after consolidation grouting.

4 Parameters of the consolidation grouting with special requirements may be determined by engineering analogy or site tests.

15.1.5 Grout materials shall be selected according to the engineering geological, hydrogeological and tunnel operation conditions. When the ground water is erosive, the anti-erosion materials shall be used, the pozzolanic Portland cement and slag Portland cement shall not be used.

15.1.6 In addition to Articles 15.1.2 to 15.1.5 of this code, the tunnel grouting shall comply with the current sector standard DL/T 5148, *Technical Specification for Cement Grouting Construction of Hydraulic Structures*.

15.2 Seepage Control and Drainage

15.2.1 The seepage control and drainage design shall adopt the engineering measures such as blocking, interception and drainage through comprehensive analysis, according to the engineering geological and hydrogeological conditions of the surrounding rock along the tunnel and the design requirements, considering the project-specific conditions.

15.2.2 Drain holes should be set in the deep-covered pressure tunnel section with external water pressure dominating the lining design. The hole spacing, row spacing and depth of the drain holes shall be determined according to the characteristics of surrounding rock and external water conditions. Drain holes should be set above the water surface for free-flow tunnel section.

15.2.3 When setting drain holes in pressure tunnels, attention shall be paid to preventing the internal water from seeping out. For tunnel sections with fissure development and fillings in the surrounding rock, flexible permeable pipes or drainage pipes wrapped with filter geotextile shall be set in the drainage holes. For tunnel sections with poor rock conditions, drainage holes should not be set.

15.2.4 Necessary seepage control measures shall be taken for the outlet section of the pressure tunnels, shallow covered tunnel sections, tunnel sections with poor geological conditions, close adjacent tunnel sections, and tunnel section crossing highway, railway and other traffic tunnels.

16 Safety Monitoring

16.1 General Requirements

16.1.1 For safety monitoring of hydraulic tunnels, necessary monitoring items shall be provided considering the tunnel function, working conditions, topographical and geological conditions, construction methods, types of support and lining, etc. Monitoring items should be designed on the principle of combining permanent monitoring with temporary monitoring, and combining instrument monitoring with patrol inspection.

16.1.2 The safety monitoring shall be provided for typical tunnel sections in one of the following cases:

1 Grade 1 hydraulic tunnel.

2 Tunnel sections using new technologies.

3 Deep-covered, high-pressure or high flow velocity tunnels.

4 Hydraulic tunnels with a diameter (or span) of more than 10 m.

5 Tunnel sections with poor engineering geological and hydrogeological conditions.

6 Tunnel sections with the tunnel alignment passing the area with important buildings, structures or with environmental protection requirements.

7 Important tunnel plug.

16.1.3 Monitoring instruments should be arranged according to the design purpose, operation conditions, and engineering geological and hydrogeological characteristics. The embedment location of instruments shall be convenient for maintenance and installation. The embedded monitoring instruments and cables shall be properly protected.

16.1.4 The construction monitoring design shall be made for tunnel sections with poor geological conditions, and the monitoring data shall be collected and the results shall be analyzed in time.

16.2 Monitoring Items and Requirements

16.2.1 Safety monitoring items of tunnel shall be determined according to the purpose of the tunnel and surrounding rock conditions. The monitoring items should include:

1 Monitoring inside the tunnel: including the deformation of surrounding rock, internal and external water pressure, stress and strain of supporting structure, flow pattern, flow rate, flow velocity, cavitation noise, aeration concentration, water surface profile, etc.; for the tunnel plug, including the displacement, joint opening between the plug and the surrounding rock, seepage pressure on the cross section of plug and longitudinal section of tunnel, plug temperature, and working stress of typical joint bars.

2 Monitoring outside the tunnel: mainly for the tunnel alignment, including the deformation and displacement of the structures outside inlet and outlet and the slopes, adit plug, and ground water table and seepage of gullies.

16.2.2 For tunnels without support or with anchor-shotcrete support, the monitoring and measurement during construction shall comply with the current national standard GB 50086, *Technical Code for Engineering of Ground Anchorages and Shotcrete Support*.

17 Operation and Maintenance

17.0.1 Technical requirements for operation of hydraulic tunnel shall be formulated according to the function of the tunnel, taking into account the natural conditions, design conditions of the structure, and data of tests and study, etc. Technical requirements shall include the operating level, discharge flow, operating condition of the gate, and mode and rate for filling and emptying.

17.0.2 Technical requirements for operation shall define the initial filling, regular emptying, inspection and maintenance of tunnel.

17.0.3 The filling flow of power tunnel shall be determined according to its arrangement, the cross-sectional size, the leakage of ground water, the change rate of water filling pressure, etc.

17.0.4 The multi-stage water filling shall be adopted in power tunnels. The head of each stage of filling should be determined according to the total pressure head, engineering geological conditions, structure type, etc., which should be 50 m to 100 m.

17.0.5 During the initial filling, the change rate of water filling pressure in power tunnels shall be strictly controlled, which should be 5 m/h to 10 m/h. The change rate may be decreased properly for tunnels with anchor-shotcrete support or plain concrete lining, and may be increased properly for steel-lined tunnels. The change rate of water filling pressure during operation period should be determined according to the monitoring results and engineering experience.

17.0.6 The steady water pressure observation shall be carried out for power tunnels during and after the initial filling. The observation duration should be more than 24 h during staged water filling and more than 48 h after water filling. The observation duration for high-pressure tunnels should be extended properly.

17.0.7 If any anomaly such as odd pressure readings, slope deformation or seepage pressure or flow occurs during water filling or steady water pressure monitoring, the water filling shall stop immediately, and the causes shall be analyzed and the measures shall be studied.

17.0.8 The regular emptying inspection should be conducted for power tunnels.

17.0.9 The change rate of water pressure shall be strictly controlled during the emptying of power tunnel, and should be 2 m/h to 4 m/h. The maximum difference between external and internal water pressure shall be less than the design external water pressure of the lining structures during emptying.

17.0.10 Necessary facilities and signs shall be provided considering the management and maintenance, such as service access, manhole, ladder, lifting hook, chainage marks inside the tunnel, and the outside marks for important sections inside the tunnel.

Appendix A Head Loss Calculation

A.0.1 Friction head loss should be calculated by the following formulae:

$$h_f = \frac{Lv^2}{C^2 R} \quad (A.0.1\text{-}1)$$

$$C = \frac{1}{n} R^{\frac{1}{6}} \quad (A.0.1\text{-}2)$$

where

h_f is the friction head loss (m);

L is the tunnel length (m);

v is the flow velocity at an actual cross section (m/s);

C is the Chezy coefficient ($m^{1/2}/s$);

R is the hydraulic radius (m);

n is the roughness coefficient, which may be taken as per Table A.0.1.

Table A.0.1 Roughness coefficient

No.	Flow surface condition of tunnel	Roughness coefficient *n*		
		Average	Maximum	Minimum
1	Unlined rock surface			
	(1) Smooth blasting	0.030	0.033	0.025
	(2) Ordinary drilling and blasting	0.038	0.045	0.030
	(3) TBM excavation	0.017	–	–
2	Cast-in-situ concrete lining with steel formwork			
	(1) Average skill	0.014	0.016	0.012
	(2) Good skill	0.013	0.014	0.012
3	Shotcreted rock surface			
	(1) Smooth blasting	0.022	0.025	0.020
	(2) Ordinary drilling and blasting	0.028	0.030	0.025
	(3) TBM excavation	0.019	–	–
4	Steel lining	0.012	0.013	0.011

A.0.2 Local head loss should be calculated by the following formula:

$$h_{\mathrm{m}}=\xi\frac{v^{2}}{2g} \tag{A.0.2-1}$$

where

h_{m} is the local head loss (m);

ξ is the local head loss coefficient, which may be taken as per Table A.0.2;

v is the average flow velocity at an actual cross section where local head loss occurs (m/s).

Table A.0.2 Local head loss coefficient ξ

No.	Position	Shape	Local head loss coefficient ξ	Remarks
1	Intake	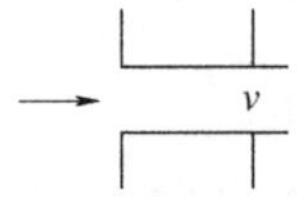 v	0.50	v is the velocity at uniform section of a pipe; r is the radius of inlet arc; d is the pipe diameter
		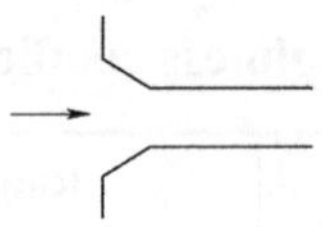	0.25	
		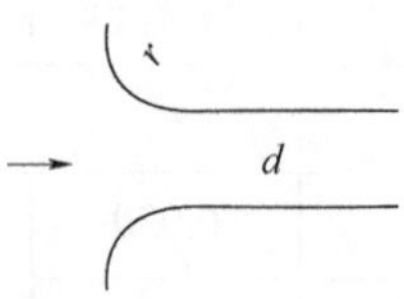 r, d	0.20 (r/d < 0.15) 0.10 (r/d ≥ 0.15)	
2	Trash rack	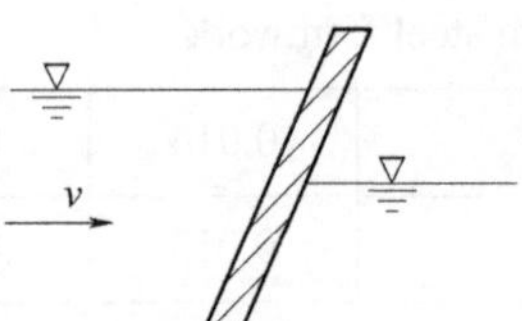 v, α	Without independent pier $\beta\left(\frac{s}{b}\right)^{4/3}\sin\alpha$	β is the bar shape coefficient; s is the bar width; b is the bar spacing; α is the inclination angle of rack; v is the average velocity in front of trash rack
3	Gate slot	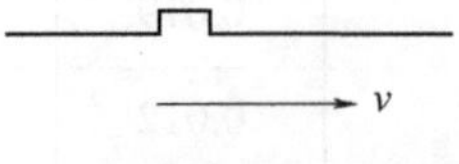 v	0.05 to 0.20	0.10 should be taken

Table A.0.2 *(continued)*

No.	Position	Shape	Local head loss coefficient ξ	Remarks
4	Rectangular tapered to circular section	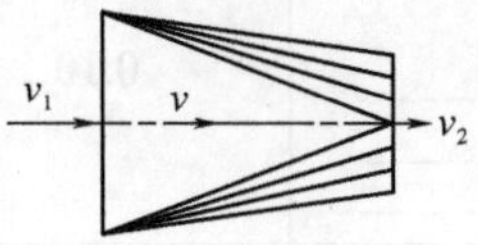 v_1 v v_2	0.05	v is the average flow velocity at transition section, $v=\frac{v_1+v_2}{2}$
5	Circular tapered to rectangular section	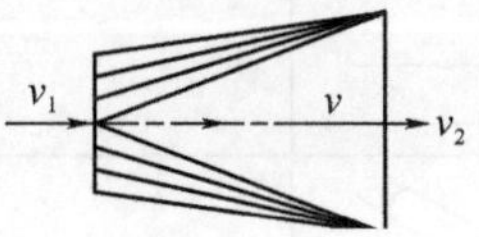 v_1 v v_2	0.10	v is the average velocity at transition section, $v=\frac{v_1+v_2}{2}$
6	Conical expansion of circular section	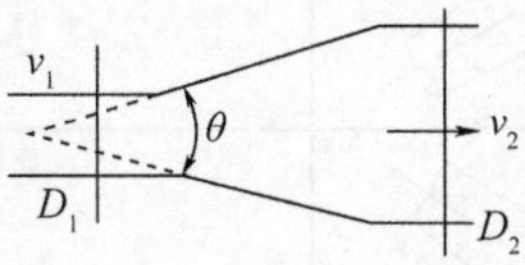 v_1 θ v_2 D_1 D_2	ξ_i see Figure A.0.2-1	v_1 is taken
7	Conical contraction of circular section	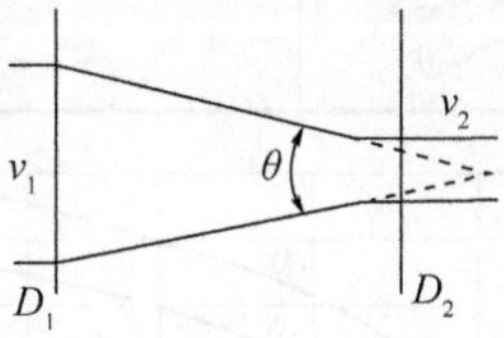 v_1 θ v_2 D_1 D_2	ξ_d see Figure A.0.2-2	v_1 is taken
8	Circular bend	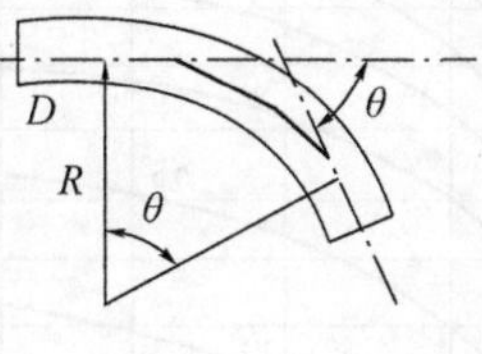 D θ R θ	$\left[0.131+0.1632\times\left(\frac{D}{R}\right)^{\frac{7}{2}}\right]\times\left(\frac{\theta}{90°}\right)^{\frac{1}{2}}$	D is the tunnel diameter; R is the bend radius; θ is the deflection angle of bend
9	Outlet	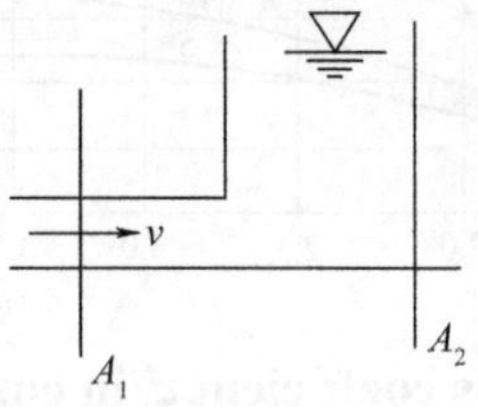 v A_1 A_2	$1-\left(\frac{A_1}{A_2}\right)^2$ 1 is taken when the downstream channel is deep	A_1 and A_2 are cross-sectional areas upstream and downstream of the outlet; v is the velocity upstream of outlet

Table A.0.2 *(continued)*

No.	Position	Shape	Local head loss coefficient ξ	Remarks
10	Right angle bifurcation		0.10	–
			1.50	–
11	Symmetrical Y-shaped bifurcation		0.75	Without conical pipe section
			0.50	With conical pipe section

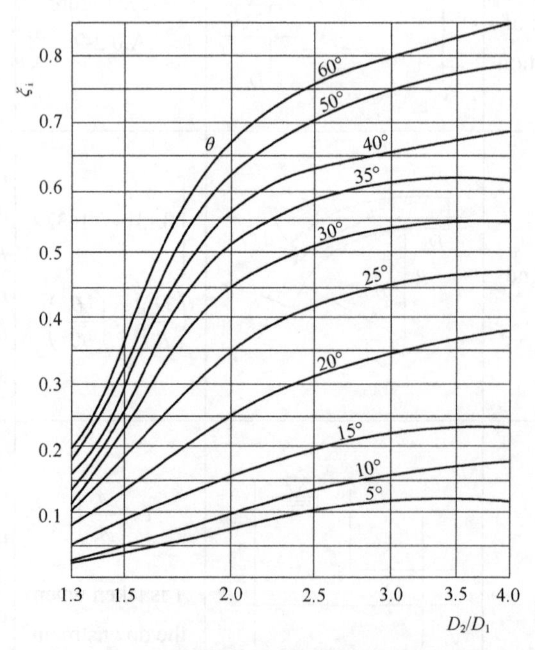

Figure A.0.2-1 Local head loss coefficient ξ_i in conical expansion of circular section ($\theta < 60°$)

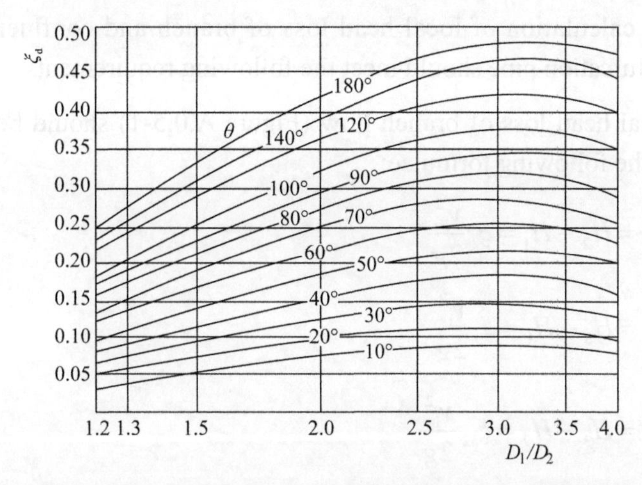

Figure A.0.2-2 Local head loss coefficient ξ_d in conical contraction of circular section

A.0.3 Bar shape coefficient β of trash rack shall be in accordance with Table A.0.3.

Table A.0.3 Bar shape coefficient β

Bar shape							
β	2.42	1.83	1.67	1.035	0.92	0.76	1.79

A.0.4 When the butterfly valve is fully opened (Figure A.0.4), the local head loss coefficient ξ shall be selected as per Table A.0.4 according to the ratio of butterfly valve thickness t to pipe diameter D.

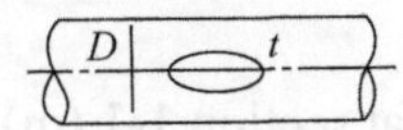

Key

D pipe diameter

t butterfly valve thickness

Figure A.0.4 Sketch of fully opened butterfly valve

Table A.0.4 Local head loss coefficient ξ

t/D	0.10	0.15	0.20	0.25
ξ	0.05 to 0.10	0.10 to 0.16	0.17 to 0.24	0.25 to 0.35

NOTE ξ may be approximately taken as 0.2, when the butterfly valve is fully opened and there is an absence of relevant data.

A.0.5 The calculation of local head loss of branch and confluence flow in T-shaped bifurcation pipe should meet the following requirements:

1 Local head loss of branch flow (Figure A.0.5-1) should be calculated by the following formulae:

$$H_{12} = H_2 - H_1 = \xi_2 \frac{v_1^2}{2g} \tag{A.0.5-1}$$

$$H_{13} = H_3 - H_1 = \xi_3 \frac{v_1^2}{2g} \tag{A.0.5-2}$$

$$H_{23} = H_3 - H_2 = \xi_{32} \frac{v_1^2}{2g} \tag{A.0.5-3}$$

$$\xi_2 = -0.95(1-q_2)^2 - q_2^2\left(1.3\cot\frac{\theta}{2} - 0.3 + \frac{0.4-0.1\psi}{\psi^2}\right)\left(1-0.9\sqrt{\frac{\rho}{\psi}}\right) - 0.4\left(1+\frac{1}{\psi}\right)(1-q_2)q_2\cot\frac{\theta}{2} \tag{A.0.5-4}$$

$$\xi_3 = -0.58q_2^2 + 0.26q_2 - 0.03 \tag{A.0.5-5}$$

$$\xi_{32} = (1-q_2)\left\{0.92 + q_2\left[0.4\left(1+\frac{1}{\psi}\right)\cot\frac{\theta}{2} - 0.72\right]\right\} + q_2^2\left[\left(1.3\cot\frac{\theta}{2} - 0.3 + \frac{0.4-0.1\psi}{\psi^2}\right)\left(1-0.9\sqrt{\frac{\rho}{\psi}}\right) - 0.35\right] \tag{A.0.5-6}$$

$$\rho = \frac{r}{D} \tag{A.0.5-7}$$

where

H_1 is the total head at section 1-1 (m);

H_2 is the total head at section 2-2 (m);

H_3 is the total head at section 3-3 (m);

H_{12} is the local head loss between sections 1-1 and 2-2 (m);

H_{13} is the local head loss between sections 1-1 and 3-3 (m);

H_{23} is the local head loss between sections 2-2 and 3-3 (m);

v_1 is the average flow velocity at section 1-1 (m/s);

θ is the intersection angle between main pipe and branch pipe (°);

ψ is the ratio of the cross-sectional areas of the main pipe to the branch pipe;

ρ is the rounding coefficient;

D is the diameter of main pipe (m);

r is the rounding radius at the intersection between branch pipe and main pipe (m);

q_2 is the ratio of branch pipe flow Q_2 to main pipe flow Q_1 before bifurcation.

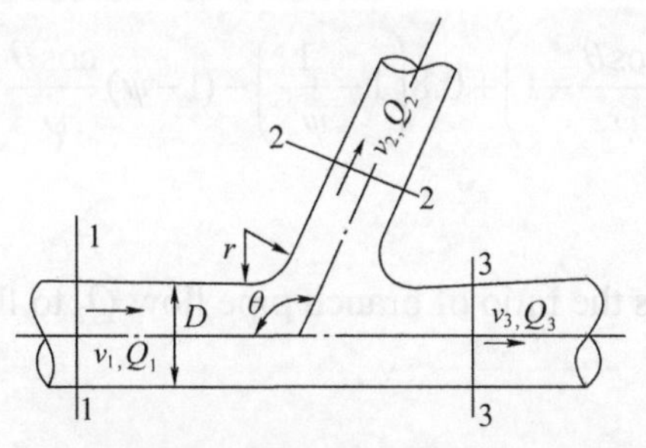

Figure A.0.5-1 Calculation sketch of local head loss of bifurcation flow

2 Local head loss of combining flow (Figure A.0.5-2) should be calculated by the following formulae:

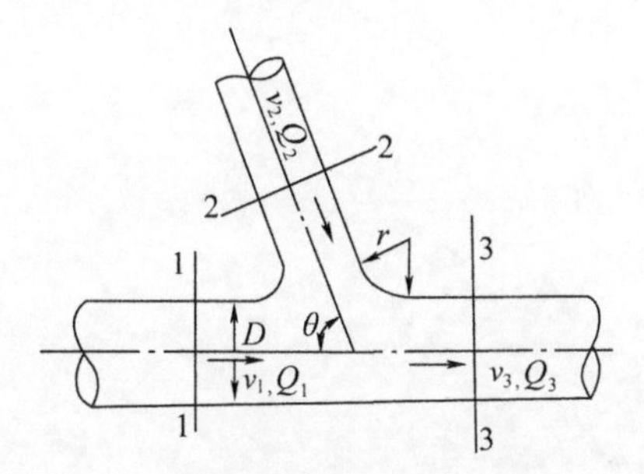

Figure A.0.5-2 Calculation sketch of local head loss of confluence flow

$$H_{12} = H_2 - H_1 = \xi_2' \frac{v_3^2}{2g} \quad \text{(A.0.5-8)}$$

$$H_{13} = H_3 - H_1 = \xi_3' \frac{v_3^2}{2g} \quad \text{(A.0.5-9)}$$

$$H_{23} = H_3 - H_2 = \xi_{32}' \frac{v_3^2}{2g} \quad \text{(A.0.5-10)}$$

$$\xi_2' = -0.95(1+q_2)^2 + q_2^2\left[1+0.42\left(\frac{\cos\theta}{\psi}-1\right)\right.$$

$$\left.-0.8\left(1-\frac{1}{\psi^2}\right)+(1-\psi)\left(\frac{\cos\theta}{\psi}-0.38\right)\right] \quad \text{(A.0.5-11)}$$

$$\xi_3' = q_2^2\left[2.59+\left(1.62-\sqrt{\rho}\right)\left(\frac{\cos\theta}{\psi}-1\right)-0.62\psi\right]$$

$$+q_2\left(1.94-\psi\right)-0.03 \quad \text{(A.0.5-12)}$$

$$\xi_{32}' = (1+q_2)\left[0.92+q_2(2.92-\psi)\right]+q_2^2\left[\left(1.2-\sqrt{\rho}\right)\right.$$

$$\left.\times\left(\frac{\cos\theta}{\psi}-1\right)+0.8\left(1-\frac{1}{\psi^2}\right)-(1-\psi)\frac{\cos\theta}{\psi}\right] \quad \text{(A.0.5-13)}$$

where

q_2 is the ratio of branch pipe flow Q_2 to flow Q_3 after confluence.

Appendix B Calculation of Discharge Capacity for Hydraulic Tunnel with Pressure Inlet

B.0.1 For a hydraulic tunnel with a pressure inlet, the pressure inlet section may be classified as short and long types. The discharge coefficient μ of short pressure inlet section may be 0.83 to 0.93; the discharge coefficient μ of long pressure inlet section may be calculated by the following formulae when the sluice gate is fully opened in free outflow:

$$\mu=\frac{1}{\sqrt{1+\left(\sum\xi_{ji}+\sum\xi_{fi}\right)\left(\frac{A_k}{A_i}\right)^2}} \tag{B.0.1-1}$$

$$A_k=Bh_2 \tag{B.0.1-2}$$

where

μ is the discharge coefficient of orifice or pipe when the gate is fully opened;

ξ_{ji} is the local head loss coefficient, including head loss coefficients of intake, transition section, gate slot and bend, etc.;

ξ_{fi} is the friction head loss;

A_k is the area of orifice (m^2);

B is the width of orifice (m);

h_2 is the height of orifice (m);

A_i is the section area corresponding to ξ_{ji} and ξ_{fi} (m^2).

B.0.2 The discharge capacity of pressure inlet tunnel may be calculated by the following formula when the sluice gate is fully opened in free outflow:

$$Q=\mu A_k\sqrt{2g\left(H_0-h_2\right)} \tag{B.0.2}$$

where

Q is the discharge (m^3/s);

g is the acceleration of gravity, taken as 9.81 m/s^2;

H_0 is the total head from upstream water level to orifice invert level, including the head of approach velocity (m).

B.0.3 The vertical contraction coefficient ε of the flat bottom sluice gate without lateral contraction may be adopted as per Table B.0.3-1; The vertical contraction coefficient ε of radial gate may be adopted as per Table B.0.3-2. The discharge coefficient of long pressure inlet section may be calculated by the following formulae when the sluice gate is partially opened in free outflow:

$$\mu_{\mathrm{j}}=\frac{1}{\sqrt{1+\left(\sum\xi_{\mathrm{j}i}+\sum\xi_{\mathrm{f}i}\right)\left(\frac{A_{\mathrm{c}}}{A_i}\right)^2}} \tag{B.0.3-1}$$

$$A_{\mathrm{c}}=B\varepsilon e \tag{B.0.3-2}$$

where

μ_{j} is the discharge coefficient of orifice or pipe when the gate is partially opened;

A_{c} is the control section area at outlet (m^2);

ε is the vertical contraction coefficient of the gate;

e is the gate opening (m).

Table B.0.3-1 Vertical contraction coefficient ε of flat gate

e/h_2	0.10	0.15	0.20	0.25	0.30	0.35	0.40	0.45	0.50	0.55	0.60	0.65	0.70	0.75
ε	0.615	0.618	0.620	0.622	0.625	0.628	0.630	0.638	0.645	0.650	0.660	0.675	0.690	0.705

Table B.0.3-2 Vertical contraction coefficient ε of radial gate

α	35	40	45	50	55	60	65	70	75	80	85	90
ε	0.789	0.766	0.742	0.720	0.698	0.678	0.662	0.646	0.635	0.627	0.622	0.620

B.0.4 When the gate of the pressure inlet section is partially opened in free outflow, the discharge capacity may be calculated by the following formula:

$$Q=\mu_{\mathrm{j}}A_{\mathrm{c}}\sqrt{2g\left(H_0-\varepsilon e\right)} \tag{B.0.4}$$

B.0.5 When the outlet of long pressure inlet section is submerged, the discharge capacity may be calculated by the following formula:

$$Q=\mu_y A\sqrt{2g\Delta H} \tag{B.0.5}$$

where

μ_y is the discharge coefficient of orifice or pipe, which is calculated by Formula (B.0.1-1), where A_{k} is replaced by A;

A is the outflow area of orifice (m^2);

ΔH is the difference between upstream and downstream heads (m).

Appendix C Aeration Water Depth Calculation

C.0.1 The self-aerated water depth of high-velocity flow may be estimated as follows considering the project-specific conditions:

1 The aerated water depth may be estimated by Hall (L.S. Hall) Formula in steep chute or flood discharge tunnel with a flow velocity greater than 30 m/s:

$$h_a = \left(1 + K\frac{v^2}{gR}\right)h \tag{C.0.1-1}$$

where

h_a is the aerated water depth (m);

h is the water depth without aeration (m);

K is the empirical coefficient depending on the properties of the flow surface material, taken as 0.004 to 0.006 for normal concrete, 0.008 to 0.012 for coarse concrete or smooth masonry, and 0.015 to 0.020 for coarse masonry or cemented stone masonry;

v is the average cross-sectional velocity without considering the effects of wave motion and aeration (m/s);

g is the acceleration of gravity, taken as 9.81 (m/s^2);

R is the hydraulic radius (m).

2 The aerated water depth of flood discharge tunnel with gentle slope, or tunnel with a flow velocity less than 30 m/s and $v^2/(gR)$ between 9.4 and 283 may be estimated by the Wang Junyong Formulae as below:

$$h_a = \frac{h}{\beta} \tag{C.0.1-2}$$

$$\beta = 0.937\left(\frac{v^2}{gR}\frac{n\sqrt{g}}{R^{1/6}}\frac{B}{h}\right)^{-0.088} \tag{C.0.1-3}$$

where

β is the water content ratio;

n is the roughness coefficient;

B is the chute width (m).

3 For a flood discharge tunnel with a small water depth and flow fluctuation, the aerated water depth may be estimated by the following formula:

$$h_{a}=\left(1+\frac{\zeta v}{100}\right)h \tag{C.0.1-4}$$

where

ζ is the correction factor, taken as 1.0 s/m to 1.4 s/m, depending on the flow velocity and section contraction; a larger value should be taken when the flow velocity is greater than 20 m/s.

C.0.2 For the flood discharge tunnel with aerated facilities, the aerated water depth shall be calculated by considering the water depth increase due to the forced aeration effect, and 5 % to 10 % of the unaerated water depth may be taken in preliminary design.

Appendix D Calculation of Circular Pressure Tunnel Lining

D.1 Lining Calculation Under Uniform Hydrostatic Pressure

D.1.1 Single-layer reinforced concrete lining may be calculated by the following methods:

1 Under uniform internal water pressure, the sectional area of single-layer steel bars may be calculated by the following formulae:

$$A_s = \frac{p_i r_i + 1000 K_0 m}{[\sigma_s]} - \frac{1000 K_0 r_i}{E_s} \quad \text{(D.1.1-1)}$$

$$m = \frac{p_i r_i}{1000 E'_c} \ln \frac{r_o}{r_i} \quad \text{(D.1.1-2)}$$

$$E'_c = 0.85 E_c \quad \text{(D.1.1-3)}$$

$$[\sigma_s] = \frac{f_y}{\psi \gamma_0 \gamma_d} \quad \text{(D.1.1-4)}$$

where

A_s is the calculation area of lining reinforcement per linear meter (mm^2);

p_i is the internal water pressure at the top of inner surface of tunnel lining (kN/m^2), taking the design value at ultimate limit state and the standard value at serviceability limit state;

r_i is the inner radius of lining (mm);

r_o is the outer radius of lining (mm);

K_0 is the unit elastic resistance coefficient of surrounding rock (N/cm^3);

E_c is the elastic modulus of concrete (N/mm^2);

E_s is the elastic modulus of steel bar (N/mm^2);

E'_c is the elastic modulus of cracked concrete (N/mm^2);

$[\sigma_s]$ is the allowable tensile stress of steel bar (N/mm^2);

f_y is the design tensile strength of steel bar (N/mm^2);

γ_0 is the importance factor of structure;

ψ is the factor of design situation.

2 The stress check of single-layer reinforcement may be conducted by the following formula:

$$\sigma_s = \frac{p_i r_i + 1000K_0 m}{A_s + \frac{1000K_0 r_i}{E_s}} \leq [\sigma_s] \tag{D.1.1-5}$$

where

σ_s is the stress of steel bar in lining with single-layer reinforcement (N/mm^2).

D.1.2 Double-layer reinforced concrete lining may be calculated by the following method:

1 Under uniform internal water pressure, the sectional area of double-layer steel bars may be calculated by the following formula:

$$A_s = \frac{p_i r_i + 1000K_0\left(m - \frac{r_i}{E_s}[\sigma_s]\right)}{[\sigma_s]\left(1 + \frac{r_i}{r_o}\right) - E_s\frac{m}{r_o}} \tag{D.1.2-1}$$

2 The stress check of double-layer reinforcement may be calculated by the following formulae:

$$\sigma_{si} = \frac{p_i r_i + \left(E_s \frac{A_{so}}{r_o} + 1000K_0\right)m}{A_{si} + A_{so}\frac{r_i}{r_o} + \frac{1000K_0 r_i}{E_s}} \leq [\sigma_s] \tag{D.1.2-2}$$

$$\sigma_{so} = \frac{(p_i r_i^2 - E_s A_{si} m)\frac{1}{r_o}}{A_{si} + A_{so}\frac{r_i}{r_o} + \frac{1000K_0 r_i}{E_s}} \leq [\sigma_s] \tag{D.1.2-3}$$

where

σ_{si} is the stress of inner ring bar in lining with double-layer reinforcement, (N/mm^2);

σ_{so} is the stress of outer ring bar in lining with double-layer reinforcement, (N/mm^2);

A_{si} is the sectional area of inner ring bar in lining per linear meter (mm^2);

A_{so} is the sectional area of outer ring bar in lining per linear meter (mm^2).

D.2 Internal Force Calculation of Lining Subjected to Vertical Pressure of Surrounding Rock, Dead Weight of Lining, and Full Water Pressure in Tunnel

D.2.1 The internal force may be calculated without accounting for the friction between the lining and the surrounding rock. The elastic resistance of surrounding rock may be distributed in the range of 270° at the bottom of the tunnel by radial action (Figure D.2.1). The distribution of elastic resistance shall meet the following requirements:

1 When $45° \leq \phi \leq 90°$, the distribution shall be as follows:

$$K\delta = -K\delta_a \cos 2\phi \tag{D.2.1-1}$$

where

ϕ is the included angle between radial section and vertical line (°);

$K\delta_a$ is the elastic resistance value at horizontal axis on resistance distribution diagram (N/m).

2 When $90° \leq \phi \leq 180°$, the distribution shall be as follows:

$$K\delta = K\delta_a \sin^2 \phi + K\delta_b \cos^2 \phi \tag{D.2.1-2}$$

where

$K\delta_b$ is the elastic resistance value at vertical axis on resistance distribution diagram (N/m).

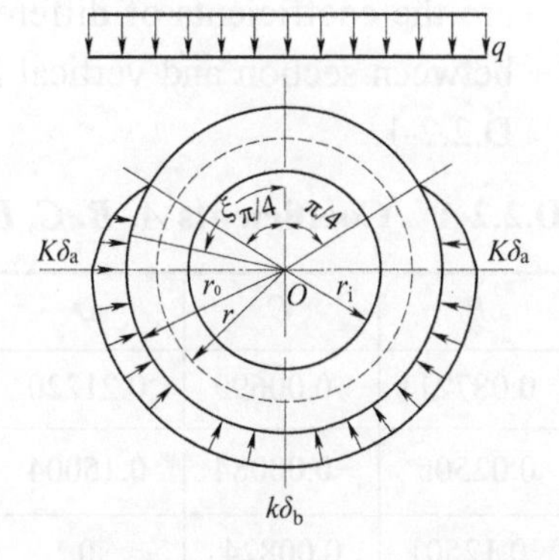

Figure D.2.1 Elastic resistance distribution at the cross section of tunnel

D.2.2 The action effects may be calculated as follows:

1 The bending moment and the axial force on each section of surrounding rock under vertical pressure may be calculated by the following formulae:

$$M = q_v r_o \, r[A\alpha + B + Cn(1+\alpha)] \quad \text{(D.2.2-1)}$$

$$N = q_v r_o \, [D\alpha + E + Fn(1+\alpha)] \quad \text{(D.2.2-2)}$$

$$\alpha = 2 - \frac{r_o}{r} \quad \text{(D.2.2-3)}$$

$$n = \frac{1}{0.06416 + \dfrac{EJ}{r^3 r_o Kb}} \quad \text{(D.2.2-4)}$$

$$K = \frac{K_0}{1000 r_o} \quad \text{(D.2.2-5)}$$

where

r	is the radius at mid-depth of the lining (mm);
q_v	is the loosening pressure of surrounding rock (N/mm^2);
M	is the bending moment on each section (N·mm);
N	is the axial force on each section (N);
K	is the elastic resistance coefficient of surrounding rock (N/mm^3);
J	is the inertia moment of lining section (mm^4);
b	is the lining width for calculation (mm);
A, B, C, D, E, F	are the coefficients of different inclination angles ϕ between section and vertical line, taken as per Table D.2.2-1.

Table D.2.2-1 Coefficients *A, B, C, D, E, F*

Section	*A*	*B*	*C*	*D*	*E*	*F*
$\phi=0$	0.16280	0.08721	−0.00699	0.21220	−0.21222	0.02098
$\phi=\pi/4$	−0.02504	0.02505	−0.00084	0.15004	0.34994	0.01484
$\phi=\pi/2$	−0.12500	−0.12501	0.00824	0	1.00000	0.00575
$\phi=3\pi/4$	0.02504	−0.02507	0.00021	−0.15005	0.90007	0.01378
$\phi=\pi$	0.08720	0.16277	−0.00837	−0.21220	0.71222	0.02237

2 The bending moment and the axial force on each section under lining dead weight may be calculated by the following formulae:

$$M = q_g r^2 (A_1 + B_1 n) \quad \text{(D.2.2-6)}$$

$$N = q_g r(C_1 + D_1 n) \quad \text{(D.2.2-7)}$$

where

q_g is the gravity of the lining section per square meter (N/mm²);

A_1, B_1, C_1, D_1 are the coefficients of different inclination angles ϕ between the section and the vertical line, taken as per Table D.2.2-2.

Table D.2.2-2 Coefficients A_1, B_1, C_1, D_1

Section	A_1	B_1	C_1	D_1
$\phi=0$	0.34477	−0.02194	−0.16669	0.06590
$\phi=\pi/4$	0.03348	−0.00264	0.43749	0.04660
$\phi=\pi/2$	−0.39272	0.02589	1.57080	0.01807
$\phi=3\pi/4$	−0.03351	0.00067	1.91869	0.04329
$\phi=\pi$	0.44059	−0.02629	1.73749	0.07024

3 Under the action of water pressure when the tunnel is full of water and there is no water head, the bending moment and axial force of each section may be calculated by the following formulae:

$$M = \gamma_w r_i^2 r(A_2 + B_2 n) \quad \text{(D.2.2-8)}$$

$$N = \gamma_w r_i^2 (C_2 + D_2 n) \quad \text{(D.2.2-9)}$$

where

γ_w is the unit weight of water (N/mm³);

A_2, B_2, C_2, D_2 are the coefficients of different inclination angle ϕ between section and vertical line, taken as per Table D.2.2-3.

Table D.2.2-3 Coefficients A_2, B_2, C_2, D_2

Section	A_2	B_2	C_2	D_2
$\phi=0$	0.17239	−0.01097	−0.58335	0.03295
$\phi=\pi/4$	0.01675	−0.00132	−0.42771	0.02330
$\phi=\pi/2$	−0.19636	0.01295	−0.21460	0.00903
$\phi=3\pi/4$	−0.01677	0.00034	−0.39419	0.02164
$\phi=\pi$	0.22030	−0.01315	−0.63126	0.03513

D.3 Calculation of Lining Subjected to Surrounding Rock Pressure, Lining Weight, Inner Full Water Pressure, and External Water Pressure

D.3.1 In the lining calculation, the resistance of surrounding rock need not be considered, but only the stratum reaction force acting on the lining semicircle and distributed radially as per the cosine law and the lateral loosening pressure of surrounding rock are taken into account.

D.3.2 The action effect may be calculated by the formulae listed in Table D.3.2-1, and shall meet the following requirements:

Table D.3.2-1 Calculation formulae of bending moment and axial force of each section

Action		M (N · mm)	N (N)
Vertical loosening pressure of surrounding rock		$q_v r_o r(A_3\alpha + B_3)$	$q_v r_o (C_3\alpha + D_3)$
Lateral loosening pressure of surrounding rock		$q_h r_o r\alpha A_4$	$q_h r_o C_4$
Lining weight		$q_g r^2 A_5$	$q_g r C_5$
Water pressure where the tunnel is full of water and there is no water head		$\gamma_w r_i^2 r A_6$	$\gamma_w r_i^2 C_6$
External water pressure	$\pi\gamma_w r_o^2 < 2(q_v r_o + \pi r q_g)$	$-\gamma_w r_o^2 r A_6$	$-\gamma_w r_o^2 C_6 + \gamma_w h_w r_o$
	$\pi\gamma_w r_o^2 \geq 2(q_v r_o + \pi r q_g)$	$\gamma_w r_o^2 r A_6 (1-2\varepsilon)$	$\gamma_w r_o^2 C_7 (1-\varepsilon) - \gamma_w r_o^2 C_6 \varepsilon + \gamma_w h_w r_o$

NOTES:

1 q_h is the lateral loosening pressure of surrounding rock (N/mm²).

2 h_w is the calculation height for uniform external water pressure (mm).

1 When the external water pressure is combined with the vertical loosening pressure of surrounding rock and the lining weight, the lateral loosening pressure of surrounding rock should be calculated by the following formula:

$$\varepsilon = \frac{2(\pi r q_g + q_v r_o)}{\pi r_o^2 \gamma_w} \tag{D.3.2-1}$$

2 When the external water pressure is combined with lining weight, the lateral loosening pressure strength of surrounding rock should be calculated by the following formula:

$$\varepsilon = \frac{2\pi r q_g}{\pi r_o^2 \gamma_w} \tag{D.3.2-2}$$

3 In Table D.3.2-1, the coefficients A_3, A_4, A_5, A_6, B_3, C_3, C_4, C_5, C_6, C_7, and D_3 of different inclination angles ϕ between section and vertical line shall be taken as per Table D.3.2-2.

Table D.3.2-2 Coefficients A_3, A_4, A_5, A_6, B_3, C_3, C_4, C_5, C_6, C_7, D_3

Coefficient	Section				
	ϕ=0	ϕ=π/4	ϕ=π/2	ϕ=3π/4	ϕ=π
A_3	0.16280	−0.02504	−0.12500	0.02505	0.08720
B_3	0.06443	0.01781	−0.09472	−0.01097	0.10951
A_4	−0.2500	0	0.25000	0	−0.25000
A_5	0.27324	0.01079	−0.29755	0.01077	0.27324
A_6	0.13662	0.00539	−0.14878	0.00539	0.13662
C_3	0.21220	0.15005	0	−0.15005	−0.21220
D_3	−0.15915	0.38747	1.00000	0.91625	0.79577
C_4	1.00000	0.50000	0	0.50000	1.00000
C_5	0	0.55535	1.57080	1.96957	2.00000
C_6	−0.50000	−0.36877	−0.21460	−0.36877	−0.50000
C_7	1.50000	1.63122	1.78540	1.63123	1.50000

D.4 Calculation of Lining Under Uniform Internal Water Pressure and Other Loads

D.4.1 Single-layer reinforced concrete lining shall be calculated as follows:

1 Under other loads, the sectional area of additional reinforcement in lining may be calculated as follows:

$$A_s' = \frac{-\sum N h_0 + 2\sum M}{2h_0[\sigma_s]} \tag{D.4.1-1}$$

2 The total sectional area of reinforcement may be calculated as follows:

$$\sum A_s = A_s + A_s' \tag{D.4.1-2}$$

3 The total sectional area of reinforcement shall not be less than that corresponding to the minimum reinforcement ratio of lining. The total

sectional area of reinforcement shall be checked by the following formula:

$$\sigma_s = \frac{p_i r_i + 1000K_0 m}{\sum A_s + \frac{1000K_0 r_i}{E_s}} + \frac{-\sum N h_0 + 2\sum M}{2h_0 \sum A_s} \leq [\sigma_s] \quad \text{(D.4.1-3)}$$

D.4.2 Double-layer reinforced concrete lining shall be calculated as follows:

1 Under other loads, the sectional area of lining additional reinforcement may be calculated by the following formulae:

$$A'_{si} = \frac{-\sum N(h_0 - a) + 2\sum M}{2(h_0 - a)[\sigma_s]} \quad \text{(D.4.2-1)}$$

$$A'_{so} = \frac{-\sum N(h_0 - a) - 2\sum M}{2(h_0 - a)[\sigma_s]} \quad \text{(D.4.2-2)}$$

2 The total sectional area of reinforcement may be calculated by the following formulae:

$$\sum A_{si} = A_{si} + A'_{si} \quad \text{(D.4.2-3)}$$

$$\sum A_{so} = A_{so} + A'_{so} \quad \text{(D.4.2-4)}$$

3 The total sectional area of reinforcement shall not be less than that corresponding to the minimum reinforcement ratio of lining. The total sectional area of reinforcement shall be checked by the following formulae:

$$\sigma_{si} = \frac{p_i r_i + (E_s \frac{\sum A_{so}}{r_o} + 1000K_0)m}{\sum A_{si} + \sum A_{so}\frac{r_i}{r_o} + \frac{1000K_0 r_i}{E_s}} + \frac{-\sum N(h_0 - a) + 2\sum M}{2(h_0 - a)\sum A_{si}} \leq [\sigma_s]$$

(D.4.2-5)

$$\sigma_{so} = \frac{(p_i r_i^2 - E_s \sum A_{si} m)\frac{1}{r_o}}{\sum A_{si} + \sum A_{so}\frac{r_i}{r_o} + \frac{1000K_0 r_i}{E_s}} + \frac{-\sum N(h_0 - a) - 2\sum M}{2(h_0 - a)\sum A_{so}} \leq [\sigma_s]$$

(D.4.2-6)

D.5 Checking Calculation of Crack Width of Lining

D.5.1 Considering uneven distribution of crack width and influence of long-term loading,the maximum crack width of tunnel lining under axial tension, large eccentric tension and large eccentric compression may be calculated by the following formulae:

$$w_{max} = 2\left(\frac{\sigma_s}{E_s}\psi - 0.7\times10^{-4}\right)l_f \quad (D.5.1\text{-}1)$$

$$\psi = 1 - \alpha_2\frac{f_{tk}}{\mu\sigma_s} \quad (D.5.1\text{-}2)$$

$$l_f = \left(60 + \alpha_1\frac{d}{\mu}\right)v \quad (D.5.1\text{-}3)$$

$$\mu = \frac{A_s}{1000H} \quad (D.5.1\text{-}4)$$

where

w_{max} is the maximum crack width (mm);

σ_s is the stress in tensile reinforcement of lining at serviceability limit state (N/mm^2);

l_f is the average spacing between cracks (mm);

ψ is the strain non-uniform coefficient of longitudinal tensile reinforcement between cracks. When $\psi < 0.3$, ψ is taken as 0.3;

α_1, α_2 are the calculation coefficients. Axial tension: α_1 is taken as 0.16, α_2 is taken as 0.60. Large-eccentricity tension: α_1 is taken as 0.075, α_2 is taken as 0.32. Large-eccentricity compression: α_1 is taken as 0.055, α_2 is taken as 0.235;

d is the diameter of tensile reinforcement (mm). For small-eccentricity tension, d is the diameter of reinforcement on the side with greater reinforcement stress;

μ is the tensile reinforcement ratio;

A_s is the total area of tensile reinforcement (mm^2). Under axial tension, it is the sum of the area of the inner and outer reinforcement. Under large-eccentricity tension, it is the area of the tension side;

H is the lining thickness (mm). It is the lining effective thickness under large-eccentricity tension;

f_{tk} is the characteristic value of axial tensile strength of concrete (N/mm^2);

v is the coefficient related to the surface shape of the tensile reinforcement, taken as 0.7 for deformed reinforcing bars, 1.0

for plain reinforcement, and 1.25 for cold drawn low carbon reinforcement wire.

D.5.2 For eccentrically compressed lining with e_0 less than $0.5H$, the checking of crack width is not required. In the case of small-eccentricity tension, the crack width may be checked according to axial tension.

Appendix E Calculation of External Water Pressure of Tunnel Concrete Lining

E.0.1 The external water pressure acting on the concrete lining structure of hydraulic tunnel may be calculated by the following formula:

$$p_o = \beta_e \gamma_w h_e \tag{E.0.1}$$

where

p_o is the external water pressure acting on the calculation point of outer surface of lining (N/mm²);

β_e is the reduction coefficient of external water pressure;

γ_w is the unit weight of water (N/mm³);

h_e is the water head from groundwater table line to the calculation point of tunnel lining (mm).

E.0.2 The reduction coefficient of external water pressure, β_e, may be taken as per Table E.0.2, considering the groundwater activity and its influence on the stability of surrounding rock, and the seepage control and drainage of tunnel.

Table E.0.2 Reduction coefficient of external water pressure β_e

Level	Groundwater activity	Influence of groundwater on stability of surrounding rock	β_e
1	Dry or moist tunnel wall	No influence	0 to 0.20
2	Seepage or drip along discontinuities	Weathering fillings in discontinuities, reducing shear strength of discontinuities, and softening soft rock mass	0.10 to 0.40
3	Large amount of dripping water, water flow in a line, or water jet along fissures or weak discontinuities	Argillizing fillings in weak discontinuities, reducing shear strength of discontinuities, and softening medium-hard rock mass	0.25 to 0.60
4	Dripping severely, with small amount of gushing water flow from weak discontinuities	Scouring fillings in discontinuities, speeding up the weathering of rock mass, softening and argillizing weakness zones such as faults, making them swell and disintegrate, causing piping, producing seepage pressure, and swelling thin weak zones	0.40 to 0.80

Table E.0.2 *(continued)*

Level	Groundwater activity	Influence of groundwater on stability of surrounding rock	β_e
5	Severe inrush water flow, and large amount of gushing water flow at weak zones, such as faults	Flushing away fillings in discontinuities, disintegrating rock mass, producing seepage pressure, swelling thick weak zones and causing collapse of surrounding rock	0.65 to 1.00

E.0.3 For hydraulic tunnels adjacent to water sources such as reservoirs, rivers or other pressure tunnels, or with complex hydrogeological conditions such as impervious curtains or drainage galleries along tunnel alignment, the seepage field should be analyzed to determine the groundwater pressure. Seepage control and drainage measures shall be taken to reduce the groundwater pressure acting on the tunnel, when the lining structure calculation is predominated by groundwater pressure.

Appendix F Stress Calculation of Circular Tunnel Lining Under Uniform External Water Pressure

F.0.1 For the circular tunnel under uniform external water pressure, the tangential stress at lining structural point may be calculated by the following formulae:

$$\sigma_{cq} = -p_o \frac{t^2}{(t^2-1)}\left(1+\frac{r_i^2}{r^2}\right) \tag{F.0.1-1}$$

$$t = \frac{r_o}{r_i} \tag{F.0.1-2}$$

where

σ_{cq} is the tangential stress at the calculation point of lining (N/mm^2);

r is the radius of lining structure at the calculation point (mm).

F.0.2 The tangential stress at the inner edge of lining may be calculated by the following formulae:

$$\sigma_c = -p_o \frac{2t^2}{(t^2-1)} \leq [\sigma_c] \tag{F.0.2-1}$$

$$[\sigma_c] = \frac{f_c}{\psi \gamma_0 \gamma_d} \tag{F.0.2-2}$$

where

σ_c is the tangential stress at the inner edge of lining (N/mm^2);

$[\sigma_c]$ is the allowable value of compressive stress of lining concrete (N/mm^2);

f_c is the design value of compressive strength of lining concrete (N/mm^2).

Explanation of Wording in This Code

1 Words used for different degrees of strictness are explained as follows in order to mark the differences in executing the requirements in this code.

1) Words denoting a very strict or mandatory requirement:

"Must" is used for affirmation, "must not" for negation.

2) Words denoting a strict requirement under normal conditions:

"Shall" is used for affirmation, "shall not" for negation.

3) Words denoting a permission of a slight choice or an indication of the most suitable choice when conditions permit:

"Should" is used for affirmation, "should not" for negation.

4) "May" is used to express the option available, sometimes with the conditional permit.

2 "Shall meet the requirements of..." or "shall comply with..." is used in this code to indicate that it is necessary to comply with the requirements stipulated in other relative standards and codes.

List of Quoted Standards

GB 50086, *Technical Code for Engineering of Ground Anchorages and Shotcrete Support*

GB 50201, *Standard for Flood Control*

GB 50287, *Code for Hydropower Engineering Geological Investigation*

GB/T 51394, *Standard for Load on Hydraulic Structures*

NB 35047, *Code for Seismic Design of Hydraulic Structures of Hydropower Project*

NB/T 35056, *Design Code for Steel Penstocks of Hydroelectric Stations*

DL/T 5057, *Design Specification for Hydraulic Concrete Structures*

DL/T 5148, *Technical Specification for Cement Grouting Construction of Hydraulic Structures*

DL/T 5166, *Design Specification for River-Bank Spillway*

DL 5180, *Classification & Design Safety Standard of Hydropower Projects*

DL/T 5207, *Technical Specification for Abrasion and Cavitation Resistance of Concrete in Hydraulic Structures*

DL/T 5398, *Design Specification for Intake of Hydropower Station*